Sensemaking in Safety Critical and Complex Situations Human Factors and Design

Sensemaking in Safety Critical and Complex Situations
Human Factors and Design

Edited by
Stig Ole Johnsen and Thomas Porathe

CRC Press
Taylor & Francis Group
Boca Raton London New York

CRC Press is an imprint of the
Taylor & Francis Group, an **informa** business

First edition published 2021
by CRC Press
6000 Broken Sound Parkway NW, Suite 300, Boca Raton, FL 33487-2742

and by CRC Press
2 Park Square, Milton Park, Abingdon, Oxon, OX14 4RN

CRC Press is an imprint of Taylor & Francis Group, LLC

Library of Congress Cataloging-in-Publication Data
Names: Porathe, Thomas, editor. | Johnsen, Stig Ole, editor.
Title: Sensemaking in safety critical and complex situations : human factors and design / edited by Thomas Porathe and Stig Ole Johnsen.
Description: Boca Raton : CRC Press, 2021. | Includes bibliographical references and index.
Identifiers: LCCN 2020043774 (print) | LCCN 2020043775 (ebook) | ISBN 9780367435189 (hbk) | ISBN 9781003003816 (ebk)
Subjects: LCSH: Human engineering.
Classification: LCC T59.7 .S46 2021 (print) | LCC T59.7 (ebook) | DDC 620.8/2—dc23
LC record available at https://lccn.loc.gov/2020043774
LC ebook record available at https://lccn.loc.gov/2020043775

ISBN: 978-0-367-43518-9 (hbk)
ISBN: 978-0-367-42243-1 (pbk)
ISBN: 978-1-003-00381-6 (ebk)

Typeset in Times
by codeMantra

Contents

Introduction

This book is a collection of practices and theory to support sensemaking. Our aim is to support working safely through having a more balanced system approach based on technology, human and organisational issues when automation is implemented. The collection highlights the importance of human factors and sensemaking in design of systems used in safety critical operations. Working safely and fulfilling, must be based on user centric design and meaningful human control in combination with our regard for new technology for a better society. We have focused on collaboration with automation and automated systems, especially in the maritime sector. Trying to learn from accident investigations, and best practices we present a set of inter-related issues. From evolution of Human Machine Interfaces, accident investigations, safety critical task analysis, sensemaking in organisations and in accident investigations, Human Factors in ship design and operations, discussion of user centric agile design, sensemaking and meaningful human control in automation, to HMI interfaces for improved sensemaking.

A short summary of all the contributions is provided below to give a broad overview of the book.

1 – INTRODUCTION, HOW HMI HAS BEEN EVOLVING

This chapter gives a short introduction to how human–machine interface (HMI) has evolved going from analogue instruments (such as gauges) local to the plant equipment to centralized computer-based control systems enabling control of complex operations. Knowledge of how to provide a performance-based system between the human operator and the computer system is required to reduce the risk of accidents. At the end of this chapter, we have a short suggestion about good practices related to HMI design and how to deliver a good return on investment (ROI) from an HMI upgrade.

2 – A GUIDE TO HUMAN FACTORS IN ACCIDENT INVESTIGATION

Human factors takes a systems approach to the understanding of behaviour, which is investigated in relation to performance shaping factors in the work environment and the identification of 'latent hazards'. Investigations often begin with a 'micro-ergonomic' analysis, focusing on events in the immediate environment when the event took place. This is followed by a 'macro-ergonomic' analysis which views the accident as the end result of a process, seeking to understand why and how latent hazards and other performance shaping factors were present when the accident occurred. The use of 'situation awareness' as an explanatory principle in accident investigation is critically analysed. Risk compensation is discussed as a psychological mechanism that can be useful in the analysis of violations. Fatigue and stress are discussed in relation to job design and accident proneness. Examples of how good design can promote or prevent error are given. Safety can be seen as a way of determining 'what happens next' through the operation of safety constraints (ranging from the design of the user interface to the design of organizational policies and procedures) providing a richer insight into accident causation, enabling safety recommendations to be made at different levels and directed to a wider range of authorities.

3 – WHAT MAKES A TASK SAFETY CRITICAL? A BARRIER-BASED AND EASY-TO-USE ROADMAP FOR DETERMINING TASK CRITICALITY

Several approaches to risk and barrier management integrate the contribution of human performance and factors by addressing safety critical tasks. As demonstrated by investigations of disasters such as the Macondo well blowout, human interaction with modern safety critical systems can be highly complex. Consequently, risk management practitioners are commonly faced with the challenge of pinpointing which tasks should be devoted the most attention and resources. As such it is often necessary to determine the level of task criticality to decide on the need for more in-depth analysis and follow-up by use improvements in design, procedures and training. When dealing with a large number of tasks, this process needs to be efficient enough to not be overly time consuming and costly. At the same time, it needs to be accurate enough so that critical tasks are not incorrectly categorized (and prioritized). This chapter discusses what characteristics defines a task as safety critical and suggests an easy-to-use roadmap for screening and identifying safety critical tasks based on their level of criticality.

4 – MAKING SENSE OF SENSEMAKING IN HIGH-RISK ORGANIZATIONS

Sensemaking – the human quest for meaning – has been an influential perspective in organization studies over the last decades. Ultimately, safety in high-risk organizations is created by the everyday behaviour of all employees in the organization – at all levels – as they go about getting their job done. Sensemaking is about how individuals and organizations make sense of or give meaning to events and experiences. Sensemaking informs and constrains identity and action and has a central role in the understanding of human behaviour. Sensemaking has been the subject of considerable research that has become scattered over several different domains with differing approaches. In relation to the extensive literature on sensemaking this chapter is a brief review of sensemaking to establish an understanding of the concept, especially how it is understood in the context of sharp-end operators in high-risk environments. This context is exemplified by the work performed on the bridge of maritime vessels, especially focusing on how the development in ship bridge design and automation influence sensemaking.

5 – PROSPECTIVE SENSEMAKING IN COMPLEX ORGANIZATIONAL DOMAINS: A CASE AND SOME REFLECTIONS

In late 2018 the Norwegian frigate *KNM Helge Ingstad* collided with an oil tanker sailing on an intersecting course. The accident occurred in an area where ship traffic was monitored by a shipping control centre. Afterwards, critical questions were asked about how such an accident could occur at all, including the publication of an audio log where one could hear the ones involved communicate. The Accident Investigation Board Norway points out that misunderstanding of the situation on the frigate was the root cause, while at the same time pointing out criticisms of the shipping control centre as well as the tanker. This chapter explores the actions of the involved parties prior to the accident focusing on communicative and coordinative practices including the use of distinct technology, as well as the configuring of events and participants' expectations. The chapter explores how one can understand this type of accident going beyond traditional sensemaking by drawing on the idea of prospective sensemaking, i.e. sensemaking processes understood analytically as primarily forward oriented. The chapter demonstrates a way to operationalize prospective sensemaking on the grounds of the KNM Helge Ingstad accident and discusses potential benefits from an operationalization of prospective sensemaking within complex organizational domains.

6 – THE CHALLENGES OF SENSEMAKING AND HUMAN FACTORS IN THE MARITIME SECTOR – EXPLORING THE *HELGE INGSTAD* ACCIDENT

The chapter discusses the challenges of sensemaking and human factors in the maritime sector based on accident and incident reports, especially the *Helge Ingstad* collision from 2018.The *Helge Ingstad* accident should not come as a surprise when looking at the poor quality of cognitive ergonomics on the bridge (i.e. poor task analysis, poor configuration of equipment, poor alarms, poor redundancy of cues, noise). The accident shows the consequences of performing piecemeal building of control systems on the bridge instead of focusing on a unified sensemaking design based on cognitive ergonomics.

7 – ADDRESSING HUMAN FACTORS IN SHIP DESIGN: SHALL WE?

Research shows that more than 80% of accidents and incidents at sea were caused by human errors and human-related factors. Less adequate design was identified as one significant factor that moderated human errors. Unlike other industries, the way human factors issues are managed in the maritime industry is not fully integrated through the life cycle of a ship, and it is still optional. "Human factors" is a relatively novel concept to naval architects and marine engineers. It covers a broad discipline with various dimensions, including habitability, workability, controllability, occupational health and safety (OHS), maintainability, manoeuvrability and survivability. From the human factor's perspective a ship can be considered as a living space, as well as a working place. A ship is usually equipped with one main control centre on the bridge, one local control centre for the engine room and more local control centres for other purposes. This chapter discusses how human factors are addressed in ship design, their implementations and challenges. Demanding issues experienced by the crew include fatigue, poor automation, flooding of alarms and imperfect design. Some challenges are reported, including inadequate standards and slowly adapting criteria, despite similar incidents occurred over and over again.

8 – SENSEMAKING IN PRACTICAL DESIGN: A NAVIGATION APP FOR FAST LEISURE BOATS

Every year a number of Norwegians are killed in small boats crashing into rocks and islands due to faulty navigation. "Human error," like inattention and misconception, is often a triggering factor. To help the human operator to make sense of a complex situation instruments like the sextant and the nautical chart have since long been developed. Recently these aides have also been automated and made electronic. However, these modern navigation tools are most often made for larger boats that can provide shelter for the sensitive equipment and many of the accidents happen with very fast open crafts and water scooters where the driver is unprotected and has to use both hands to hold on and steer the boat. This chapter describes a sensemaking-in-practice project, the design concept, development and test procedure of an application specially tailored for such fast, open and exposed boats. The smart phone-based application was developed according to human-centred design (HCD) principles in a project named SikkerKurs and a proof-of-concept was tested at sea in Norway in 2017 with good results. The user experience was the focal point for the user tests. Local users have for some time been using the application and delivered critique and comments. A second iteration is now being planned for the autumn 2020.

9 – UNIFIED BRIDGE – DESIGN CONCEPTS AND RESULTS

This chapter examines the effect of how a user-centred design process was a differentiator when designing a new ship bridge environment. The concept development started by involving the user from the ideation phase and throughout the product development process towards a finished product. The final product was released into the market as a ship bridge environment designed with a holistic perspective including a redesign and re-arrangement of the physical consoles, input devices and software interfaces located in the environment to support four design criteria: safety, simplicity, performance and proximity. To fulfil the criteria, a methodological approach including insight studies, operator interviews and eye-tracking was selected. The continuous feedback loop from the crew onboard the vessel using the ship bridge in daily operation was important to gain further insight and continue improving the concept. After 5 years in the market a benchmark insight study was carried out and the feedback from the vessel operators was undoubtedly positive, with reluctance to return to a more conventional ship bridge environment.

10 – SUPPORTING CONSISTENT DESIGN AND SENSEMAKING ACROSS SHIP BRIDGE EQUIPMENT THROUGH OPEN INNOVATION

Ship bridges are complex work environments outfitted with a variety of operational equipment typically supplied by different vendors. A lack of standardized design elements creates a wide variety of interface design characteristics once all systems are integrated together within a single working environment. Thus, seafarers must engage with, and manage, a plethora of incongruent human–machine interfaces on a ship's bridge during operations. Inconsistencies across bridge systems can negatively affect seafarer sensemaking whilst increasing cognitive demands and likelihood for error. Systemic challenges within the maritime industry enable suboptimal design processes and outcomes that ultimately impact seafarers and sensemaking at the sharp end of operations. This chapter presents an initiative working towards a more integrated approach for multi-vendor maritime equipment design through open innovation. We address current industry gaps by developing dynamic interface guidelines based on web technologies, user-centred design principles and open-source component libraries. An ECDIS (Electronic Chart Display and Information System) design case is presented to illustrate the application of the design guideline and its contribution for enhancing sensemaking in ship operations.

11 – USER-CENTRED AGILE DEVELOPMENT TO SUPPORT SENSEMAKING

User-centred design (UCD) is a fitting approach for supporting sensemaking, as the aim of UCD is to understand contextual user needs and use this as a basis to iteratively explore and design solutions. Similarly, agile software development uses iterations and rapid feedback to continuously explore, learn and improve user stories and their implementations. The user-centred and agile approaches represent mindsets of great influence in their respective fields of design and software development. In this chapter, we will briefly introduce the two approaches. We present five models that can be used to combine agile and user-centred development and discuss three common challenges in user-centred agile development (UCAD). Key take aways from the chapter are:

- The importance of exploring, understanding and framing the problem, in order to solve the right challenge, propose solutions that fit the context of use, make sure user needs drives ideation – not the love of technology.
- The importance of user involvement and contextual testing as a team effort to enable interdisciplinary decision-making and collaboration, enable continuous learning and improvement based on feedback, create solutions that work in real-life scenarios and contexts.

12 – IMPROVING SAFETY BY LEARNING FROM AUTOMATION IN TRANSPORT SYSTEMS WITH A FOCUS ON SENSEMAKING AND MEANINGFUL HUMAN CONTROL

The chapter discusses the safety and security challenges of autonomous industrial transport systems. The chapter covers all four transportation domains (road, rail, air and sea), common important concepts and how the various domains can learn from each other. Suggested key measures related to organizational, technical and human issues are presented focusing on the importance of involving humans in the loop during design and operations, support sensemaking, focus on learning from projects through data gathering and risk-based regulation. Unanticipated deviations are key challenges in automated systems, together with how to design for human–automation interaction and meaningful user involvement.

13 – APPLICATION OF SENSEMAKING: DATA/FRAME MODEL, TO UAS AIB REPORTS CAN INCREASE UAS GCS RESILIENCE TO HUMAN FACTOR AND ERGONOMICS SHORTFALLS

Unmanned Aircraft Systems (UAS) have grown exponentially. The operators remotely control the entire UAS flight from the Ground Control Station (GCS) while sitting comfortably hundreds or even thousands of miles away. Its pilot safety feature drove UAS' initial development towards security, law enforcement and military. However, it also led to the elimination of standardized testing required for the manned aircrafts. The hastily developed and deployed UAS lead to an increased number of UAS mishaps. As studies show 69% of all mishaps are due to human factors proliferation in GCS, and nearly 25% of those mishaps are directly related to human factors and ergonomic (HF/E) shortfalls in GCS design. Nowadays, UAS are being employed in several sectors. Nonetheless, UAS-specific standards and methodical testing are still lacking. This study verifies the applicability of existing ANSI/HFES-100 for computer workstation to UAS GCS, followed by a case study to apply sensemaking (data/frame model) to UAS Accident Investigation Report (AIB) to identify HF/E shortfalls in GCS that may have been overlooked previously. Once the HF/E shortfalls are identified, a human factor standard ANSI/HFES-100 is used to resolve those issues in GCS retroactively, thus improving UAS GCS resilience.

14 – CONSTRAINED AUTONOMY FOR A BETTER HUMAN–AUTOMATION INTERFACE

Most industrial autonomous systems use an operator to handle situations beyond the automation system's capabilities. This means regular changes between automatic control and human control. A safe change from automatic to human control requires, among other factors, that humans have enough time from being alerted to the problem, until getting sufficient situational awareness to act safely, i.e. a *maximum response time*. This chapter explores this problem from the point of view of the automation system and will provide a framework for describing and analysing it. This is based on the definition of an *operational envelope*, which defines what the control system, including humans and automation, needs to be able to handle in the different system states, to achieve the system objectives. By observing constraints on how automatic functions are implemented, it is possible to divide the operational envelope into distinct areas based on the definition of a *response deadline*. This is the minimum time a human has available to respond to any new situation that requires human intervention. When response deadline is longer than the maximum response time, it is safe to leave control to the automation and rely on alerts to muster the human.

15 – HMI MEASURES FOR IMPROVED SENSEMAKING IN DYNAMIC POSITIONING OPERATIONS

The drive towards increasing levels of automation is a major force for improving safety across all complex, safety-critical industries. This has not only led to impressive safety records but also to new types of accidents where the collaboration between human operators and automated systems breaks down. In this chapter we look at this challenge from the perspective of a modern ship bridge and more specifically the operation of Dynamic Positioning systems (DP systems). Accidents and near-misses reports indicate that the sensemaking of Dynamic Positioning Operators (DPOs) is not always successful, and we have identified six main sensemaking-related challenges: (1) alarms, (2) mode surprises, (3) critical information hidden from view, (4) "Private" Human Machine Interfaces (HMIs) limits shared situation awareness, (5) deskilling and (6) out-of-the loop. This chapter looks at these challenges from the perspective of making HMI improvements and introduces the concept of overview displays. We discuss the feasibility of utilizing such displays to make safety improvements based on lessons learned from the petroleum and nuclear domain as well as a user-centred concept study performed with experienced DPOs.

Contributions

Miriam Eileen Nes Begnum
NTNU - Norwegian University of
    Science and Technology Gjøvik,
    Norway

Frøy Birte Bjørneseth
NTNU - Norwegian University
    of Science and Technology &
    Kongsberg Maritime Ålesund,
    Norway

Robert S. Bridger
Knowledge Sharing Events Ltd.
President of the Chartered Institute of
    Ergonomics and Human Factors.
Lee-on-the-Solent, UK.

Brit-Eli Danielsen
NTNU - Norwegian University of
    Science and Technology Trondheim,
    Norway

Tor Erik Evjemo
SINTEF Trondheim, Norway

Kay Fjørtoft
SINTEF Ocean Trondheim, Norway

Åsa Snilstveit Hoem
SINTEF & NTNU - Norwegian
    University of Science and
    Technology Oslo, Norway

Lars Hurlen
IFE - Institute for Energy Technology
Halden, Norway

Gunnar Jenssen
SINTEF Trondheim, Norway

Stig O. Johnsen
SINTEF Trondheim, Norway

Steven C. Mallam
USN - University of South-Eastern
    Norway Vestfold, Norway

Terje Moen
SINTEF Trondheim, Norway

Ian Nimmo
User Centered Design Services, INC.
    Arizona, USA

Kjetil Nordby
The Oslo School of Architecture and
    Design Olso, Norway

Thomas Porathe
NTNU - Norwegian University of
    Science and Technology Trondheim,
    Norway

Ørnulf Jan Rødseth
SINTEF Ocean Trondheim, Norway

Vincentius Rumawas
WEsustain.no AS, Asker, Norway

Qaisar R. Waraich
Smartronix, Inc.
The George Washington University
    Maryland, USA

Sondre Øie
DNV Ålesund, Norway

Contributors

Dr Miriam Eileen Nes Begnum has a PhD in Computer Science from the Norwegian University of Science and Technology (NTNU) and has 15 years of experience in the field of user-centred and inclusive design. She started her career as an IT consultant, did R&D work at MediaLT 2007–2012 and from there went on to the Institute for Design, NTNU Gjøvik where she was responsible for the Master Programs in User-Centered Media Design and Interaction Design from 2012 to 2014 and nominated by her students for Study Quality Award twice. Currently, Begnum is a Senior Advisor at the Norwegian Welfare and Labour Administration (NAV), Section of Design, where she is supporting agile teams on universal design and promoting inclusive service design.

Dr Frøy Birte Bjørneseth is an Associate Professor in Maritime Human Factors in the Department of Ocean Operations and Civil Engineering at the Norwegian University of Science and Technology in Ålesund, Norway. She has a PhD in Computer Science with focus on human–machine interaction and human factors within the maritime environment. She is also a Principal Engineer in Human Factors and Control Centres with 13 years of experience from Kongsberg Maritime CM AS, a worldwide supplier of maritime equipment. Frøy Birte's research focuses on human factors in demanding maritime operations and holistic user-centred design thinking in ship bridge and control centre environments, including design and evaluations of software applications and physical equipment.

Dr Robert S. Bridger is an independent consultant, writer and public speaker in Human Factors. He served as Head of the Human Factors Department at the Royal Navy's Institute of Naval Medicine for 19 years. Before joining INM, he ran the postgraduate program in ergonomics at the University of Cape Town. He is author/co-author of over 200 research papers and reports, sole author of the text-book *Introduction to Human Factors and Ergonomics*, now in its fourth edition, and author of the new book *A Guide to Active Working in the Modern Office*. Dr Bridger has extensive experience applying human factors to accident investigations in the maritime domain and, more recently, has been working with the Health Safety Investigation Branch of the National Health Service in the United Kingdom. He is President of the Chartered Institute of Ergonomics and Human Factors.

Brit-Eli Danielsen has over the last 12 years worked at CIRiS – Centre for Interdisciplinary Research in Space, Trondheim, where research performed on the International Space Station has been central. The human factors interest has been connected to work in the CIRiS control room for space operations as control room operator, training manager and as project manager of design and development of control room. Danielsen is currently pursuing a PhD in Industrial Design, focusing on ship bridge design.

Dr Tor Erik Evjemo has a master's degree and a PhD in sociology from the Norwegian University of Science and Technology, NTNU. He is a researcher at SINTEF where research interests are work, organization and society with a special interest in organizational, technological and social aspects of safety, mainly in the sectors of oil and gas, health, as well as transport and aviation in particular. He is particularly interested in how qualitative methods are used in the field of safety science.

Kay Fjørtoft, MSc, is a Senior Researcher at SINTEF Ocean. He has a broad research background and has been working with maritime research for more than 20 years, with topics such as software architecture and development, autonomy, integrated operations, maritime safety operations, Arctic, freight transport, port community systems and communication (telecom) including sensor technology. Kay has participated in writing several books, papers and articles, which focuses on software architecture and logistics, as well as maritime operational challenges.

Åsa Snilstveit Hoem is a PhD student at the Department of Design at the Norwegian University of Science and Technology (NTNU). She completed her MSc in Reliability, Availability, Maintainability and Safety (RAMS) at the same university in 2014 and has worked as a researcher at SINTEF Community, Department of Mobility and Economics, and at the Department of Safety Research from 2014. The topic of the PhD project is risk analysis in the design phase of Maritime Autonomous Surface Ships (MASS). Through her PhD work she has been involved in various projects related to autonomous transportation and joined the Human and Autonomy Lab (HAL) at Duke University.

Lars Hurlen has a degree in Mechanical Engineering from NTNU in Norway. After working a few years as an interaction designer and internet consultant he started at Institute for Energy Technology (IFE) where he now holds the position of senior scientist. During his 13 years at IFE, Lars has been involved with a number of large-scale industry projects, simulator-based research experiments and early stage conceptual studies, investigating various aspects of well-functioning human–machine interaction with an emphasis on innovation in interface design, information visualization and control room design.

Dr Gunnar Jenssen is a Human Factors expert with a PhD in Transport Engineering on Behavioral Adaptation and Safety Effects of Advanced Driver Assistance Systems and a MSc in Psychology. As Senior Research Scientist at SINTEF Transport Research, since 1988 he has extended his PhD work to user acceptance of completely autonomous vehicles and feasibility of autonomous winter maintenance at airports.

Dr Stig O. Johnsen is Senior Researcher at SINTEF in Norway. He has a PhD from NTNU in Norway with a focus on resilience in complex socio-technical systems and has a Master of Technology Management from MIT/NTNU. He is chairing the Human Factors in Control network (HFC) in Norway to strengthen the Human

Factors focus during development and implementation of safety-critical technology. His research interest has been meaningful human control to support safety and resilience during automation and digitalization.

Dr Steven C. Mallam is an Associate Professor at the Faculty of Technology, Natural Sciences and Maritime Sciences, University of South-Eastern Norway. He obtained his PhD in Human Factors from Chalmers University of Technology (Sweden) and Master of Science specializing in Ergonomics from Memorial University of Newfoundland (Canada). Steven's primary research interests involve the analysis and optimization of end-users and their work processes within complex socio-technical systems through improved knowledge mobilization, training and design.

Terje Moen has a BSc in the field of Electronics and has been Senior Advisor at SINTEF Transport Research since 1990. His areas of expertise are research methodologies with emphasis on instrumentation, data acquisition and analysis and high focus of advanced vehicle technology, driver assistance systems and cooperative intelligent transport systems (C-ITS).

Ian Nimmo has a degree in Electrical Engineering from the UK, 10 years as the Process Controls Engineering Manager for Imperial Chemical Industries Ltd. in the UK at Teesside Operations, over 50 years of experience in the processing industry. He was involved in all the relevant standards and guidelines and is working for Honeywell as a Senior Engineering Fellow researching Abnormal Operations as the Program Director for the ASM™[1] Consortium for ten years. Currently, he is the President and CEO of User Centred Design Services Inc. for over 20 years (recently Inducted into the Process Control Magazines Hall of Fame).

Dr Kjetil Nordby is a Professor in Interaction Design with the Institute for Design at The Oslo School of Architecture and Design and leader of the Ocean Industries Concept Lab. He holds a master's in Interaction Design from Umeå Design School and a PhD from the Oslo School of Architecture and Design. Kjetil has extensive experience working as project leader, researcher, interaction and industrial designer for industry and academia.

Sondre Øie is a Principal Consultant in DNV GL with 12 years of experience of providing technical advice and management consultancy to clients in various high-risk industries, including petroleum, rail and hydro power. His expertise within Human Factors and risk management is built on top of a MSc degree in organizational psychology from the Norwegian University of Science and Technology. Sondre has been working mostly with offshore safety and major accident prevention, particularly within the domain of oil well drilling, focusing on well control and integrity. In the recent years Sondre has increased his focus towards the maritime industry and is now primarily working with projects involving the design and operation of autonomous

---

[1] ASM™ is a trademark of Honeywell IAC·

and remotely operated vessels. In addition to delivering projects for clients Sondre is frequently involved in authoring industry guidelines and standards, participating or leading R&D projects, and submitting and presenting papers for various safety, risk and reliability conferences.

Dr Thomas Porathe has a degree in Information Design from Malardalen University in Sweden. He is currently Professor of Interaction Design at the Norwegian University of Science and Technology in Trondheim, Norway. He is specialising in maritime human factors and design of maritime information systems, specifically directed towards control room design, e-navigation and autonomous ships. He has been working with e-Navigation since 2006 in EU projects like BLAST, EfficienSea, MONALISA, ACCSEAS, SESAME and the unmanned ship project MUNIN. He is active in the International Association of Aids to Navigation and Lighthouse Authorities (IALA).

Ørnulf Jan Rødseth has an MSc in electronic engineering from 1983. He is a well-known researcher for more than 25 years in maritime information and communication technology. In the last years, he has worked mainly with autonomous ship technology and maritime digitalization. He is Senior Scientist at SINTEF Ocean and is the general manager in Norwegian Forum for Autonomous Ships. He is a member of ISO TC8 and IEC TC80 and regularly meets at IMO as observer for ISO.

Dr Vincentius Rumawas earned his bachelor's degree (BEng) in Offshore Engineering from the Faculty of Marine Technology, Surabaya Institute of Technology (ITS), Indonesia. He also studied industrial psychology and holds a Bachelor of Arts (BA) degree from the University of Airlangga, Surabaya, and a Master of Science (MSc) degree in Educational Psychology from the University of 17 Agustus 1945, Surabaya. Rumawas had been working as a Lecturer and Researcher in the Faculty of Marine Technology, ITS, for 13 years before he continued for his master's degree (MSc) in the Department of Marine Technology, NTNU, specializing in marine structures. Rumawas pursued his study and gained his PhD from the same university. His research topic was Human Factors in Marine Design which lies within the corridor of risk and reliability engineering. Rumawas' previous research and projects cover a wide range of topics from structural engineering, sea transportation, port management, logistics and supply chain management, safety in shipping, to human- and organization-related problems. Currently, Rumawas is working as a Senior Working Environment and Human Factors Consultant in WE sustain.no.

Dr Qaisar R. Waraich has a PhD in Systems Engineering from George Washington University, Washington, D.C. He has more than 17 years of experience working with numerous U.S. Department of Defense (DoD) unmanned aircraft systems (UAS). He has participated in research with National Aeronautics and Space Administration (NASA). He has worked with American National Standards Institute (ANSI) for the development of UAS Ground Control Station

(GCS) specific standards. He has been actively involved in organizing and judging the Association for Unmanned Vehicle Systems International (AUVSI) Small UAS (SUAS) competition for more than 12 years. He has authored technical publications and spoken at conferences. He serves as a technical reviewer for several engineering journals. He also manages a team of engineers to ensure ergonomic design for human–machine interface (HMI).

1 Introduction, How HMI Has Been Evolving

I. Nimmo
User Centered Design Services

CONTENTS

This section gives a short introduction on how human–machine interfaces (HMI) have evolved: from analogue instruments (such as gauges) local to the plant equipment to centralized computer-based control systems enabling control of complex operations. Knowledge of how to provide a performance-based design between the human operator and the computer system is required to reduce the risk of accidents. At the end of this chapter, a few suggestions on good practices related to HMI design and how to deliver a good Return On Investment (ROI) from an HMI upgrade are provided.

IN THE BEGINNING, HMI EVOLVED AS TECHNOLOGY BECAME AVAILABLE

HMI have been evolving for the last 50 years. In the early days of instrumentation, the HMI was local to the instrument, for example, a pressure gauge, a temperature bulb, a level indicator. As instrument signals extended over greater distances to the local control room, local control room indications were provided and collected together onto an instrument panel. These were still gauges, but some were collected into charts. Some signals now provided process values (PVs) to pneumatic controllers and resulted in PID (proportional–integral–derivative) control of valves and other devices (Figure 1.1).

A PID controller is a control loop using feedback that is widely used in industrial control systems. A PID controller continuously calculates a deviation as the difference between the desired setpoint and a measured process variable and applies an error correction based on proportional, integral, and derivative algorithms.

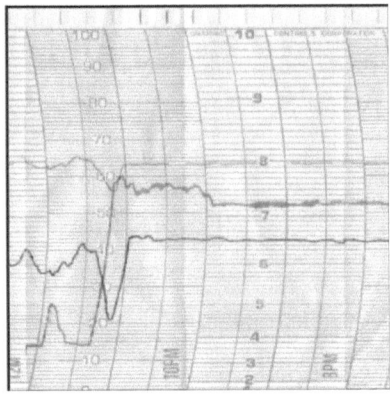

FIGURE 1.1   Pneumatic controller.

FIRST, COMPUTER INTERFACES WERE MERE COPIES OF THE HARDWARE SOLUTIONS

Technology progressed, and instrumentation became a microprocessor with new capabilities and transmission over greater distances. They were eventually ending up connected to a computer through input and output cards. Signals became a mixture of analogue and digital. The visual display unit (VDU) became the host of the instrumentation displays, often copied from their pneumatic or electronic representation.

The pneumatic controls became obsolete and were replaced by individual electronic controls but in the same configuration as the pneumatic controls on a metal panel. This solution had many limitations due to space availability and transmission distance for the signal, which often was still pneumatic (Figure 1.2).

A "current to pressure" transducer (I/P) converts an analogue signal to a proportional linear pneumatic output. The I/P converter provides a reliable means of converting an electrical signal into pneumatic pressure in many control systems. The "pressure to current" converters (P/I) converts a pneumatic signal to a proportional electrical output.

In the past, the HMI design of hardwired panels focused on optimizing the controls based on process adjustments to allow the operator to stand at one spot on the board and make moves to control the process. As computerized control became available, the very first version provided a direct copy of these controls with optimum grouping, as described below (Figure 1.3).

NO HUMAN FACTORS OR ERGONOMICS WERE CONSIDERATIONS IN THE DESIGN

As the VDU evolved, it initially emulated the panel instrument displays. However, this became challenging as hundreds of pages of group displays (with eight single equivalent controllers) became available as more instrumentation was added to the process plants. There is a need to access relevant controllers based on alarm

FIGURE 1.2   1950s pneumatic control panels.

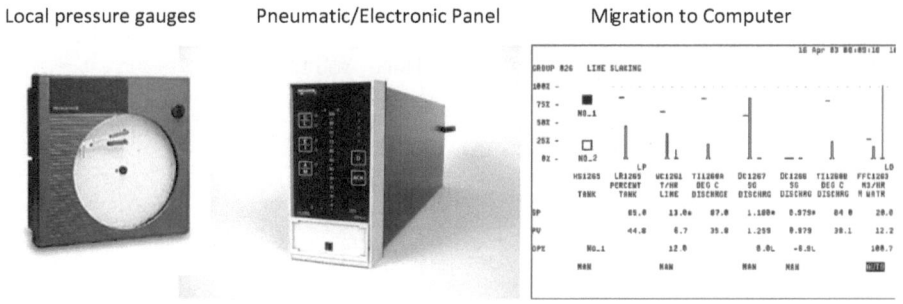

FIGURE 1.3   Evolution of controllers.

conditions. The operator keyboard shown in Figure 1.4 is designed to ease navigation to these groups, providing a quick indication of the pages with active alarms through function keys with alarm lights.

However, this gave you a key-hole view of the process and no total overview (Figure 1.5).

As technology became available and had more capability, a new graphic HMI presentation evolved, allowing more information to be added and displayed: The move from a page of 8 controller groups to new process display pictures on a black background using multiple colours to represent equipment and instrumentation values.

The industry did not spend any time deciding what to use for these displays. They observed engineers using two document types to interpret the process: a process flow diagram (PFDs) document that provided an overview of the process and on a more detailed level, the Process and Instrument Drawing (P&ID), including more details than a PFD documenting major and minor flows, control loops and instrumentation which became the main control system view of the plant.

Engineers quickly latched onto the P&ID and PFD diagram and broke them up into individual screens. Operators initially did not like them as they lost the optimization of controls being grouped on a page. They now experienced difficulty and navigation issues as operators had to find controllers on hundreds of P&ID pages and not grouped based on tasks as the previous design.

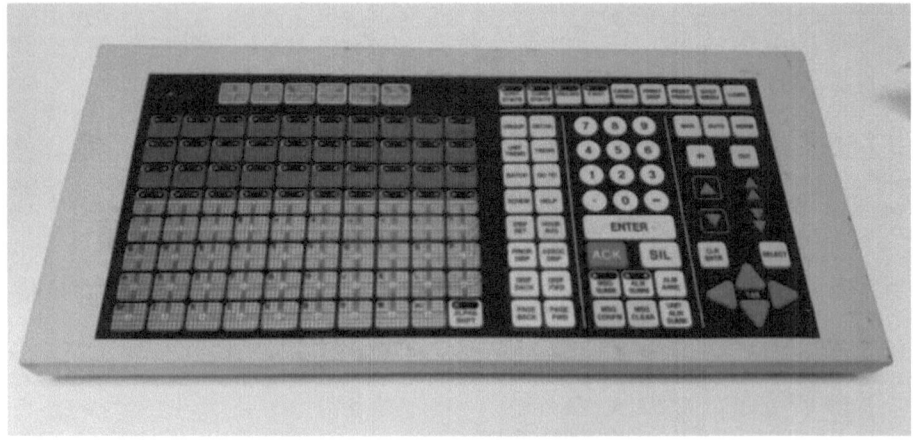

FIGURE 1.4   Operator keyboard. (Courtesy of Honeywell IAC.)

This created a key-hole view

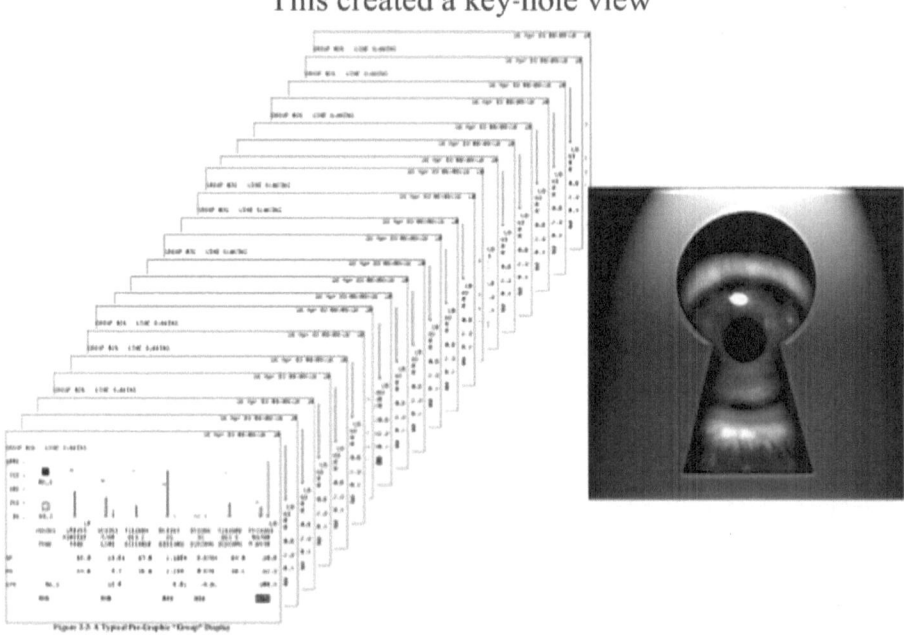

FIGURE 1.5   Hundreds of controller pages.

As the computer capabilities allowed, the HMI became a representation of the P&ID drawings, as shown in Figure 1.6, often creating many pages and a navigation nightmare (Figure 1.8). The screens were often colourful, with no human factor considerations.

Because of the lack of human factor understanding, many mistakes that are well known today were common in the designs introduced in the early 1980s. For

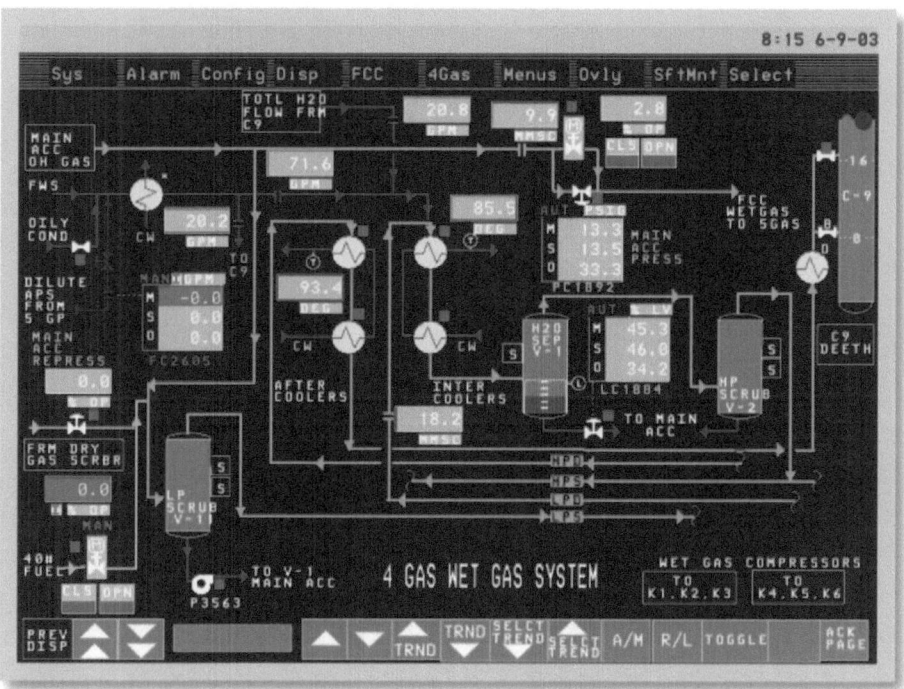

FIGURE 1.6 Process and instrument drawing (P&ID) colourful display screen.

example, today, it is well known that people with a colour deficiency (colour blind) cannot discriminate between the colours red and green. These two colours were dominant in the HMI design and differentiated open (green) versus closed (red) and running (green) versus stopped (red). Colour is a coding system, and this HMI design produced coding systems that were compromised due to human limitations.

For example, the colour "Red" was used to represent the most critical high priority alarm colour and should remain reserved for this purpose on a graphic. However, as just discussed, it sometimes indicated the status of the valve positions and motor de-energized, and many other equipment states.

Colour is just one type of coding method. I have witnessed some graphics using "Red" for over 13 different codes, rendering it no longer a sensible coding system.

As an example of more acceptable coding, from a recent HMI project, we used line thickness instead of colour to portray different electrical line voltages for an electrical transmission centre overview display.

Lack of human factors and ergonomics knowledge also led to poor readability and lack of display graphics clarity. Incorrect viewing angles and distance from the user to the screen were challenges that were often ignored making readability and noticing a change impossible. In addition, the use of too small fonts, wrong font style, and too much information squeezed onto a display also hindered readability. Thus, sometimes viewing text and numbers contributed to poor display quality as seen in Figure 1.7.

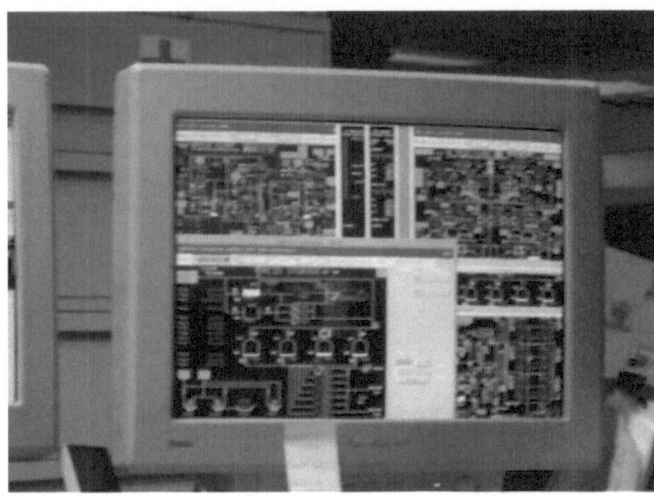

FIGURE 1.7   Computer display screen with multiple P&ID)screens open on a single display.

LACK OF PHILOSOPHY AND STYLE GUIDES

The lack of philosophy and style guides allowed individuals to do what they thought was right, rather than following a consistent design; the inconsistent placement of objects and links prevents humans from using pattern recognition skills. There could be irregular objects represented by multiple object libraries, causing an incompatible and contradictory representation of similar measurements.

Many companies allowed the evolution to continue without any rules or guidance; if specifications did exist, they ended up on a shelf and were never applied. Internal engineers changed, and new staff (not receiving any advice or direction) used the knowledge that they brought with them to design or modify existing screens, which caused a further mismatch of codes and standards.

Lack of human factors knowledge led to misleading placement of information – instruments illustrated in one place in a graphic were located in a completely different location in real life, thus causing wrong assumptions and poor decisions during the diagnosis of problems.

Today's practices dictate that a whole site and often the entire company should follow a <u>philosophy</u> which dictates consistency between different systems and technologies, <u>a Style Guide</u> provides details of codes, colours, text, font types and objects to be consistent throughout the design. These standards must be enforced and audited by the site.

The International and local Standards should enforce the rules, as shown in the Standard and Guideline Hierarchy below, i.e. statutory and regulatory requirements, and standards such as ISA 101 and ISO 11064 should be a sound basis for the corporate guidelines and specifications and the HMI philosophy (with style guidelines) (Figure 1.8).

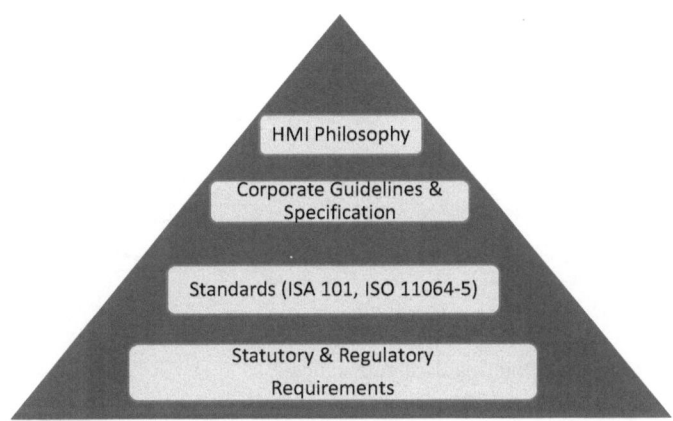

FIGURE 1.8 Standard and Guideline Hierarchy.

THE LACK OF A HIERARCHY REQUIRED MANY SCREENS TO PROVIDE AN OVERVIEW

P&IDs represent how a plant is built but not how it operates. Thus, one of the downsides of using the P&ID design representation is that navigation of information may become complicated and hard.

Figure 1.9 shows the navigation paths to retrieve just four pieces of information. However, task analysis can reveal information grouping and better ways of presenting data in the graphics.

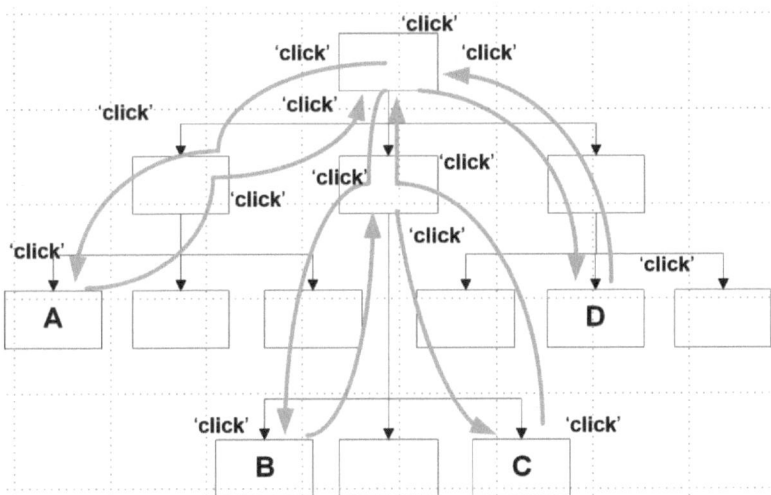

FIGURE 1.9 Navigation to screens to make moves for a task.

Test		Results	Reference Range	Indicator		
				LOW	NORMAL	HIGH
HCT	=	31.7 %	24.0−45.0			
HGB	=	10.2 g/dl	8.0−15.0			
MCHC	=	32.2 g/dl	30.0−36.9			
WBC	=	9.2 x10⁹/L	5.0−18.9			
GRANS	=	6.5 x10⁹/L	2.5−12.5			
%GRANS	=	71 %				
L/M	=	2.7 x10⁹/L	1.5−7.8			
%L/M	=	29 %				
PLT	=	310 x10⁹/L	175−500			

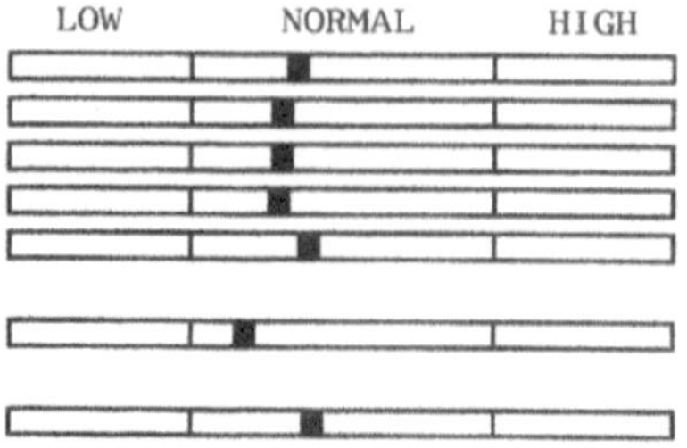

FIGURE 1.10   Example of putting data into context.

Early designs of displays also revealed numerous repeats of the same information. The task analysis provides insight into rationalizing the required data and information, offering a dramatic reduction in the total number of displays. Task analysis has typically reduced 12 current screens used to achieve a simple processing task and consolidate the required data onto one screen.

A goal of the task analysis is to consolidate data and turn data into information, as seen in the example presented in Figure 1.10.

The left side shows the usual format of data on display without any context. It relies on the operator's understanding and ability to remember 1,000s of data points, what their working range is, and then put the data in context. Short-term memory issues compromised situation awareness leading to human error and missed warnings. Having the reference range next to the value is useful, but not very practical as it requires a lot of cognitive work to make the comparison of data to the range specified. Using "reveal information" when hovering the cursor over the value is one way to achieve this goal but again not very practical. A simple graphical indicator can put the data into context immediately for the operator. We follow this practice by adding the number when the data is off-normal.

THE POWER OF THE COMPUTER ALLOWS AN INCREASE IN SCOPE OF CONTROL

The operator usually has a formal desk known as a console with screens to show each of the operating units. Unfortunately, with more and more units added, there comes the point when more screens are not practical. It is not unusual to see a console with 12–16 screens, and the operator tries to build an overview of each of the units under their scope of control.

There are some concerns and challenges that must be addressed:

- How many units can a single operator manage?
- What happens if more than one of the plant units is in a disturbance at the same time? Do they prioritize one and ignore the other.
- Does the operator have sufficient time to respond to alarms before the consequences happens?

When addressing these concerns, a systematic analysis must be performed based on a task analysis of the work that the operator is doing within the actual timeframe. A time plot of the tasks, workload analysis, and Management of Organizational Change Study are required to identify whether the operator can safely bring all those units to a safe state in a timely manner.

The analysis must also consider the number of deviations that can happen, i.e., the need for a close look at the number of alarms generated by each of the process units, which may exceed the operator's ability.

Finally, we must go back to the number of screens an operator can effectively manage. The standards suggest four screens. However, that does not allow for a second or third operator at the console during start-up and disturbances. A full Ergonomic Assessment is necessary for the control room and the desk to eliminate things that could compromise situation awareness.

CONTROL ROOMS AND SA HAVE <u>NOT</u> NECESSARILY EVOLVED WITH THE NEW HMI

A few things that can compromise situation awareness (SA) have already been mentioned. Still, a more in-depth look would require the review of the control room itself, the operator's console design, the number of screens, distance of the screen from the operator, and viewing angles. (In addition to a description of work processes, collaboration and interaction with actors outside the control room).

The screen layouts and navigation techniques must conform with the Philosophy and Style Guide, which will have rules for colour usage, text size, font types, line usage, arrow usage, navigation, hierarchy, and consistent use of objects and their placement on screens.

Once we have control of the number of screens that are optimal for the control operator, we can consider the console furniture and if features such as sit/stand will be available as recommended by most ergonomic studies. The trend today is to provide a large overview screen either above the console or on an adjacent wall. An overview of the operator's scope of control is on this display. It should have all the critical information which the operator must respond to immediately, whether it is safety, environmental, or process related. How the data should be displayed is found in the ISA 101 standard (human–machine interfaces).

Level 1 screens give an overview, where the structure can go down to level 4 details. The console desktop screens display Level 2, 3, and 4 displays. Under "Normal" circumstances just Level 2 displays would populate the desktop screens. Level 2 displays monitor and control approximately 80% of a section of the Level 1 displays.

Level 3 displays are typically the traditional P&ID view of the process with more detail than the Level 2 screens. Level 4 screens are used for fine details and diagnostics and testing information. It is still allowed to have a screen of faceplate displays similar to the original eight used in the early days, as this allows the operators to make multiple quick changes grouped by task. Hence, the hierarchy reveals more and more specifics as the operator navigates from Level 2 to 4 data (Figure 1.11).

One of the differences between traditional displays and high-performance displays is to make the operator proactive rather than the traditional being reactive. At the same

FIGURE 1.11   Control room operators consoles.

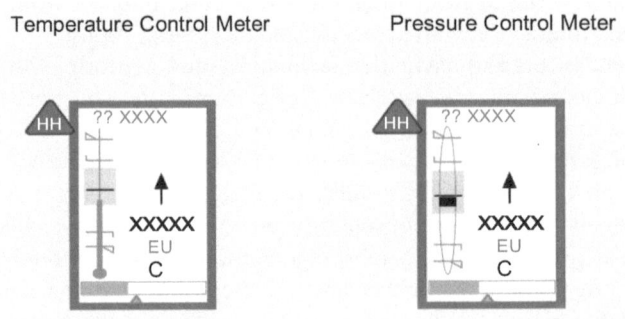

FIGURE 1.12   Instrument objects.

time, operators wait for alarms before responding to a change; the high-performance graphics have the operator identifying "Off-normal" and reacting before the process moves into the "abnormal" zone where the alarms are initiated. We do this using objects to represent flows, pressures, temperatures, and level controls.

We show the normal operating ranges on the objects and then throw a number next to or above the object if the normal band is exceeded (off-normal).

The objects in Figure 1.12 represent a temperature controller and a pressure controller, and the objects are shown as if they are in a high priority alarm condition. The shaded grey area on either side of the setpoint line represents "Normal Operating Range." The "xxxxx" represents the Process Value (PV), which can be turned on all the time or thrown on the object when the PV is "off-normal."

A big question many readers may be asking is, "how does this apply to me and my industry?" Well, regardless of your industry, all the principles stated here apply. The common factors are people with their unique skill sets, human limitations and computers with all their complexity and quarks. Having an overview is an essential practice, one that was lost in the early days of the introduction of computer technology.

Incomplete information on human factors and ergonomics has led to inadequate coding systems, poor navigation, unsuitable display formats, poor font types, and size choices. Also, improper management of change is allowing inconsistencies in the design.

CURRENT GOOD PRACTICES FOR DESIGNING HMI

The critical foundation to design a high-performance HMI is to base the design on a thorough task analysis, i.e. what tasks need to be done by the operator?

The tasks and the criticality of the tasks dictate the information requirements. The information requirements determine the HMI design.

The HMI design should be guided by a high-performance philosophy and style guide ensuring proper use of colours/content, layout, hierarchy/navigation, alarm support and the ability of the operator to be supported by several systems/sources/cues to ensure a high level of situational awareness and sensemaking. (Some of these issues are described in the *High-Performance HMI Handbook*; Hollifield et al., 2008).

The design should be driven by a user-centric design, involving user needs and perception to ensure that all the tasks are supported.

Guidelines and standards such as EEMUA 201 Control Rooms: A Guide to Their Specification, Design, Commissioning and Operation, ISO 11064 Ergonomic Design of Control Centres and ISA 101 Human–Machine Interfaces should be used. The ISA101.01 is a good standard because of its main message: keep process information display to the operator simple and based on task requirements.

Alarm guidelines such as EEMUA 191 Alarm systems, IEC 62682 Alarm Management (coming from the ISA 18.2 alarm standard) and YA-711 Principles for Design of Alarm Systems support these standards.

On a more specific level related to HMI, the standard **ISO 9241** is a multi-part standard from the International Organization for Standardization (**ISO**) covering ergonomics of human–computer interaction. It gives a broad context for the development of needed systems. High-Performance HMI aims to improve the operator's ability to "Detect" a problem, i.e., process parameters going off-normal and heading to an alarm condition. "Diagnose," providing sufficient information together to allow the operator to diagnose the problem promptly. Finally, the operator's ability to "Respond" correctly to this problem and to correct it while still in "Normal" and before the "Abnormal" or alarm condition.

REFERENCES

EEMUA 191 (2013) *Alarm systems—A guide to design, management and procurement*, Edition 3, Engineering Equipment and Materials Users Association. London, UK.

EEMUA Publication 201 (2019) *Control rooms: A guide to their specification, design, commissioning and operation*, Engineering Equipment and Materials Users Association. London, UK.

Hollifield, B., Habibi, E., Nimmo, I., & Oliver, D. (2008) *The high performance HMI handbook: A comprehensive guide to designing, implementing and maintaining effective HMIs for industrial plant operations.* Plant Automation Services, Houston, TX.

IEC-62682 (2014) *Management of alarms systems for the process industries.* IEC.

ISO 11064 (2000). *Ergonomic design of control centres (multipart standard issued from 1999 to 2020).* ISO.

ISO 9241(2019). *Ergonomics of human-computer interaction (multipart standard issued from 1993 to 2020).* ISO.

ISA-101.01 (2015) *Human machine interfaces for process automation systems.* ISA.

YA-711 (2001) *Principles for design of alarm systems.* Petroleum Safety Authority Norway.

2 A Guide to Human Factors in Accident Investigation

R. S. Bridger
Knowledge Sharing Events Ltd. Past President Chartered Institute of Ergonomics and Human Factors

CONTENTS

INTRODUCTION

This chapter summarises a presentation made by the author at a safety conference in Norway (Bridger 2015). 'Learning and changing from investigations' was the theme of the conference, and the presentation aimed to illustrate how human factors (HF) can contribute to organisational learning and, in particular, the necessity of taking a systems approach to generate safety recommendations.

It is widely recognised that HF competencies must be present in all incident investigations and incident reports. Investigators and operators must be trained in HF and/or supported by qualified HF professionals. The author is currently working with the Royal Navy on the development of a learning pathway, taking a tiered approach to train 4% of over 30,000 personnel as HF facilitators, 2% as HF supervisors, with smaller numbers of more highly trained personnel as HF trainer assessors and leads. External oversight, level 3 assurance, accreditation and mentoring will be delivered by the Chartered Institute of Ergonomics and Human Factors (CIEHF). The institute has already accredited a HF learning pathway for the energy sector. There is also interest from the National Health Service in the UK, which has a Health Safety Investigation Branch (HSIB) employing Chartered Members of the Institute and aspirations to deliver further CIEHF-approved training for HSIB employees.

TAKING A SYSTEMS APPROACH

HF is not about people. It is about systems and the processes used to design systems, taking into account the 'human factors' that place constraints on the design while providing opportunities for innovation. Systems always have a human element, no matter how automated they are, and HF is best applied when systems are designed so that they are safe to operate. This is not to say that human behaviour falls outside the scope of HF – far from it, what falls within the scope of HF is human behaviour in the context of the system. The Nobel Prize winning economist Herbert Simon offered a 'scissors analogy' (Figure 2.1) that is relevant to the application of HF in research and practice. The mind and the world fit together like the blades of a pair of scissors. If a pair of scissors does not cut well, there is no point in looking at only one of the blades (human operators). Rather, one has to look at both blades and how they fit together. That is, to understand why people behave as they do and why errors sometimes have adverse effects, it is necessary to understand the interactions and interrelations between operators and the rest of the system. These interactions are shaped both by events at the time and by decisions made earlier when the system was designed.

Understanding the context of use of a system is particularly important when safety occurrences are investigated. Investigators should focus not only on what people were doing but on the equipment they were using and the overall context, including the workload.

Safety occurrences include accidents, near misses and other events reported via the safety management system of the organisation. Such events might not have led to adverse outcomes or loss but are deemed to be worthy of further investigation and include violations of standard operating procedures and errors.

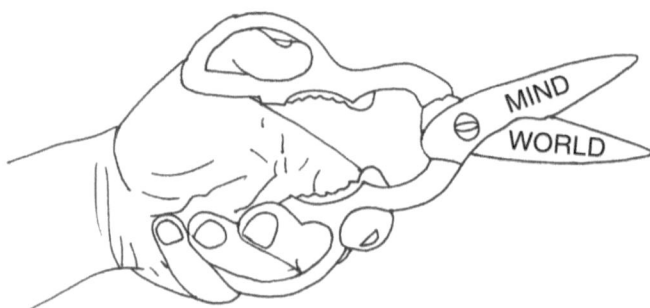

FIGURE 2.1 Human rational behaviour is shaped by a scissors whose blades are the structure of the task environments and the computational capabilities of the actor. (Herbert Simon, 'Invariants of human behaviour'. *Annual Review of Psychology*, 41, 1990, pp. 1–19.)

HUMAN FACTORS AND 'SENSE-MAKING' IN ACCIDENT INVESTIGATION

Kilskar et al. (2019) concluded that the term sense-making is often underspecified in the literature, although when it is used, the focus appears to be on operational safety, in the main, and on those involved at the time. This is in contrast to HF, which is a mature discipline applied throughout the system lifecycle (see Ministry of Defence 2015, for example). From the perspective of HF, then, questions about sense-making can be asked at any of the stages of the CADMID cycle (Conception, Design, Assessment, Manufacture, In-Use, Decommission).

In order to makes sense of safety occurrences, investigators should distinguish between operational safety and system safety. Table 2.1 (Bridger 2018) contrasts system safety and operational safety – both of which can be improved using HF analyses and design methods. In practice, what may appear to be violations of standard operating procedures –operators deliberately taking shortcuts or disabling alarms – may really reflect adaptive ways of coping with design deficiencies (in order that an unsafe or inefficient system can be operated as safely and efficiently as possible) that were not understood or properly considered during the design and assessment stages. For example, poor design of procedures or equipment may have resulted from inadequate task analysis and poor understanding of what to do in a crisis. In other words,

TABLE 2.1
Contrasting System Safety and Operational Safety

		System Safe to Operate?	
		Yes	**No**
Operated Safely?	**Yes**	**True accident**: A ship is hit by a freak wave that could not have been predicted. The event was not foreseeable. No mitigation was possible	**Technical failure**: Personnel were operating the system correctly. Design faults led to component failure (e.g. a pressure relief valve malfunctioned) or unpredictable behaviour of an automated subsystem lead to the event.
	No	**Violations or errors**: The ship ran aground because of poor communication between an understaffed and poorly trained group of officers on the bridge. Deviations from standard operating procedures and danger signs were communicated but ignored or misunderstood	**System-induced error**: The automatic system defaulted to a rarely used operating mode that was not displayed. The operators misinterpreted the behaviour of the system and failed to recognise the warning signs due to poor training and supervision.

Source: From, Bridger, RS, *Introduction to Human Factors and Ergonomics*, 4th Edition. CRC Press, Boca Raton, FL.

looking back from the apparent violation might reveal a failure to integrate HF into the systems engineering processes used to build the system in the first place. Thus, to understand these issues, HF knowledge and competence should be a part of the investigation teams.

Safety investigations are conducted as part of organisational learning in a 'just culture', the purpose of which is to generate safety recommendations and NOT to apportion blame or to obtain evidence against individuals. The latter requires a different approach, is not part of HF and is usually carried out by other authorities with different qualifications and training under a different system of governance. Learning from incidents is a topic in its own right and further information can be found in Drupsteen and Guldenmund (2014) and in the White Paper published by the Chartered Institute of Ergonomics and Human Factors (2020).

MITIGATING SOURCES OF RISK IN SYSTEMS THROUGH HUMAN FACTORS IN DESIGN

HF regards people as components of systems and not as separate entities. Technology has become increasingly reliable and is normally low risk. Safety occurrences are often traced to the actions of human operators who may even be blamed for the occurrence, whereas it is at the interfaces between system components that much of the risk normally lies. New components may be incompatible with older ones and vice versa and some components may be incompatible with operators. Attributing safety occurrences to 'human error' may oversimplify a more complicated picture as is summarised in Table 2.2. The contemporary view is that human error is a consequence of deeper issues with the system (combination of issues such as poor design, poor training, mental overload, fatigue (Dekker 2004)).

Human error is ubiquitous – *people make mistakes all the time but normally nothing happens* (Bridger et al. 2013). Such errors often go unrecorded although many high-risk industries have 'near-miss' reporting systems in an attempt to improve the resilience of the system by better understanding the kinds of errors and high-risk situations that do occur. The perspective of HF, then, is not to stop people from making

TABLE 2.2
Sources of Risk in Systems – Simplistic Interpretations

	Safety Occurrence?	
Human error?	Yes	No
Yes	A	B
No	C	D

Source: From Bridger, RS, *Introduction to Human Factors and Ergonomics*, 4th Edition. CRC Press, Boca Raton, FL.
A, Blame the operator; B, An 'unknown unknown'; C, blame something else; and D, taken for granted.

errors but to identify and eliminate error-provoking factors in the work environment that may have adverse effects.

PERFORMANCE SHAPING FACTORS (PSFs) IMPACTING HUMAN BEHAVIOUR

As we have seen, the approach of HF is to consider human behaviour in the context of the system in which people are working. Performance shaping factors (PSFs) are aspects of the work environment and of human behaviour which influence system performance. The context is nothing more than the sum of PSFs in workplace at the time a safety occurrence took place (Figure 2.2). PSFs are not causes, but they are always relevant. Good 'detective work' may be needed to identify the PSFs so that we can learn from them and improve the work environment (see, for example, Gould et al. 2006). When investigating safety occurrences, the immediate focus is on the events during a thin slice of time at the critical moment to understand the PSFs in the immediate environment and to search for latent hazards – this might be called *a micro-ergonomic analysis*. Next, the focus is on the organisation and the events leading up to the accident, seeking to discover *why* and *how* the PSFs were present in the

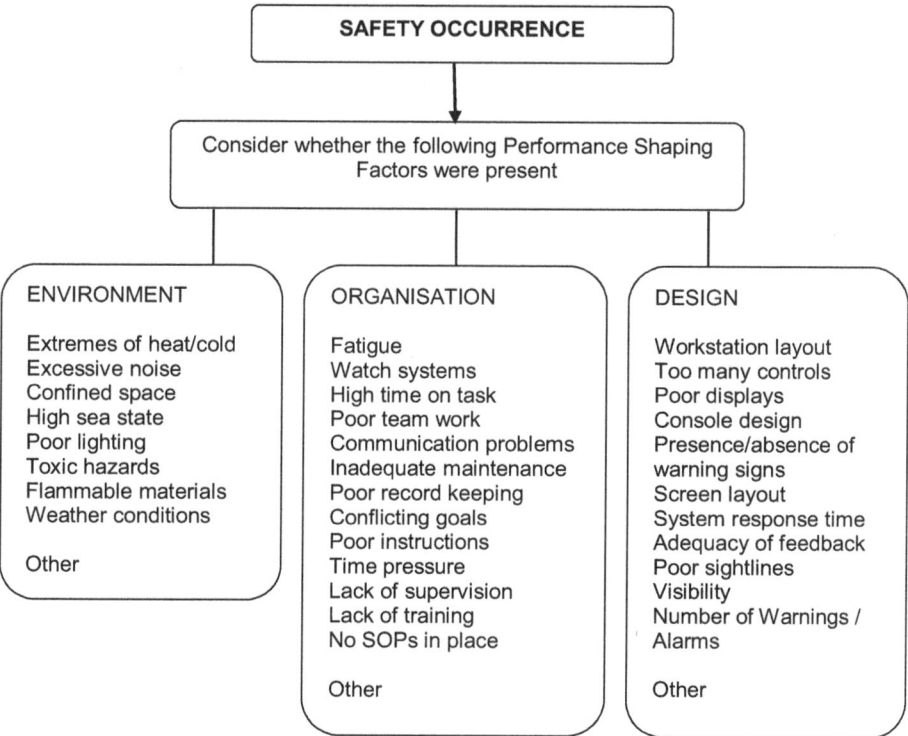

FIGURE 2.2 Performance shaping factors in safety management and accident investigation. (From Bridger et al., 2012, Crown Copyright, Contains public sector information licensed under the Open Government Licence v2.0.)

FIGURE 2.3 Conflicting information on a 2.15 kg tin of green olives may prompt erroneous purchasing decisions.

environment – *a macro-ergonomic analysis* that focuses on the safety management system and the policies and procedures embedded within it.

Latent hazards and defects. A hazard is something that can cause adverse effects – loud noise can cause hearing loss, excessive loads can cause back injury, a ladder is a hazard, because the user might fall from it, badly designed interfaces may make errors more likely or may make the consequences of an error more serious.

A risk is the likelihood that a hazard will cause an adverse effect. Likelihood can be expressed as a numerical probability ('there is a one in a million chance of the employee being injured at work') or in qualitative terms ('the risk of injury is negligible'). A *hazard* can be regarded as being *latent* when it is not extant and can only cause an adverse effect when certain PSFs are present. Figure 2.3 shows a large can of olives with a latent design defect on the label. The can contains green olives and the word 'GREEN' is written in large BLACK letters. The olives surrounding the words are BLACK olives. Latent defects of this kind often only become apparent after an adverse event has occurred (in this case, a can of green olives is purchased when the customer intended to purchase black olives).

WHAT IS SAFETY – A HUMAN FACTORS PERSPECTIVE

Passive definitions of safety focus on the state of being 'safe' – of being protected from harm or other non-desirable outcomes. More active definitions consider the essence of system safety to be the ability to control *what happens next* particularly to control recognised hazards in order to achieve an acceptable level of risk.

Systems are dynamic in operation and one way of controlling 'what happens next' is at the design stage by controlling and limiting what *can happen* next (limiting the number of states and the pathways from one state to another). Leveson (2016) sees safety as a problem of control and emphasises the importance of *safety constraints* at different levels (Figure 2.4):

- Low level controls include, for example, fuses, relief valves and other engineering solutions
- Intermediate level controls include, for example, thermostats and devices that monitor or limit aspects of the process. *Forcing functions* (Norman 2013) are excellent examples of safety constraints that limit *what can happen next* during a sequence of human actions in a system.
- Higher levels of control include, for example, standard operating procedures, maintenance programmes and training programmes
- The highest levels of control include, for example, safety committees, wider policy and legal frameworks, national and international standards and legislation.

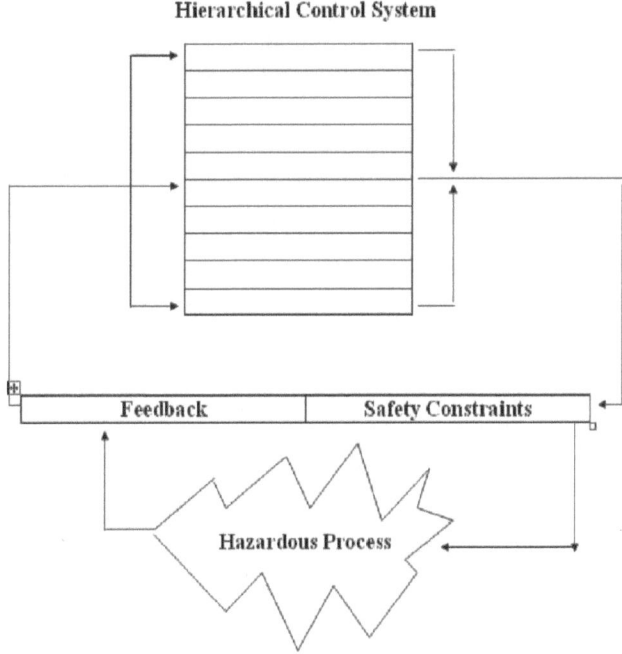

FIGURE 2.4 Safety as matter of control – a visual summary of some key ideas from Leveson (2011). There is a hierarchy of control with different levels – the particular controls and the hierarchy will depend on the organisation. (From Bridger RS, *Introduction to Human Factors and Ergonomics*, 4th Ed, CRC Press, Boca Raton, FL.)

As can be seen, safety constraints are both components of system safety (they are part of the hardware and software when the system is designed) and at a higher level of control, they are embedded in the safety management system of the organisation.

ACCIDENTS AS FAILURES OF SAFETY CONSTRAINTS

Following Leveson, if safety is a matter of control through the operation of safety constraints, accidents can be understood as a failure in the operation of one or more safety constraints, as in the following examples:

- *A safety constraint(s) was missing*: a worker repairing a factory roof slips and falls to his death because he was not wearing a safety harness.
- *A safety constraint(s) was misapplied*: A 'STOP' sign on the road at a busy intersection is misspelled 'SOTP'. A driver is distracted by the spelling mistake, overshoots the intersection resulting in a collision
- *A safety constraint(s) was not applied*: The ferry *Herald of Free Enterprise* sank in 1989 after leaving Zeebrugge resulting in the deaths of 189 people. The bow doors had not been closed before departure resulting in flooding of the lower decks.
- *A safety constraint(s) could not be activated*: At one point during the nuclear incident at the Three Mile Island nuclear power station in 1979, a pressure relief valve was stuck open, whereas a light on the control panel indicated that the valve was *closed*. In fact, the light did not indicate the position of the valve, only the status of the solenoid being powered or not, thus giving false evidence of a closed valve. As a result, the operators did not correctly diagnose the problem for several hours.
- *A safety constraint(s) was applied too late.* Gould et al. (2006) report that a Norwegian offshore patrol vessel hit a submerged rock after an inexperienced navigator lost control over his exact position. Failing to observe a waypoint, the vessel was late turning and hit a submerged rock. This was a knowledge-based error caused by a lack of training in the identification of waypoints.
- *A safety constraint(s) was removed too soon or applied for too long*: On October 31, 2014, at 1007:32 Pacific daylight time, Virgin Galactic's SpaceShipTwo (SS2) operated by Scaled Composites LLC, broke up into multiple pieces during a rocket-powered test flight. SS2 had been released from its launch vehicle, White Knight Two (WK2), N348MS, about 13 seconds before the structural breakup. Scaled was operating SS2 under an experimental permit issued by the Federal Aviation Administration's (FAA) Office of Commercial Space Transportation. After release from WK2 at an altitude of about 46,400 ft, SS2 entered the boost phase of flight. Although the co-pilot made the required 0.8 Mach callout at the correct point in the flight, he incorrectly unlocked the feather immediately afterwards instead of waiting until Space Ship Two reached the required speed of 1.4 Mach
- *A safety constraint(s) was unsafe.* A Norwegian coastguard vessel grounded when the retractable sonar dome was damaged when the vessel entered shallow waters (Gould et al. 2006). The sonar indicator was only visible

from one side of the bridge leaving the navigator unaware of the vessel's depth. The navigator could not see the indicator from his position because of the design of the bridge and failed to retract the dome when the vessel entered shallow water.

- *A safety constraint was ignored.* Prior to leaving the Port of Zeebrugge, the boatswain – the last person on G Deck of the *Herald of Free enterprise* – said that he did not close the bow doors of the ferry because it was not his duty.
- *A safety constraint was inadequate.* Mosenkis (1994) gives the example of a contact lens package that contained two, almost identical, bottles: one containing a caustic lens cleaning fluid and the other containing saline to wash the fluid off when the lens had been cleaned. Small, difficult to read labels were the only clues as to the contents of the bottles. Eye damage occurred when users cleaned the lenses with saline and then washed them with cleaning fluid. The example illustrates the failure to take into account the users' capacity throughout the task – at the critical point, the user's vision is at its poorest. Shape and colour could be used to distinguish the two bottles (e.g. a red, hexagonal bottle for the cleaner and a blue, smooth one for the saline).

In practice, when safety occurrences are being investigated, the identification of failures such as those described above is part of the initial micro-ergonomic analysis of the event. It is followed by asking deeper questions about how the organisation manages safety: how effective the constraints were; why the constraints failed and whether similar accidents had happened before (and if they had, how the organisation had responded, if at all). Answers to these questions might be found in the organisation's 'Hazard log'. A Hazard Log is defined in Def Stan 00-56 Issue 4 as: 'The continually updated record of the Hazards, accident sequences and accidents associated with a system. It includes information documenting risk management for each Hazard and Accident'.

In a safe system, a hazardous process is controlled by:

- A hierarchical control system
- Safety constraints
- Feedback about the operation of the process and the operation of the constraint (Figure 2.4).

So, when investigating an accident, we can begin by asking the following questions:

1. Were all the appropriate constraints in place?
2. Was the process controlled correctly?
3. Was the correct feedback received?

ACCIDENTS AND HUMAN ERROR AS A STARTING POINT IN ANALYSIS

Accidents are often attributed to human error. Statements such as '90% of accidents are due to human error' are common, but they are wrong. As we saw in Table 2.2, people make errors all the time and normally nothing happens – when errors result in safety occurrences, this is a sign that changes are needed, such as better documentation, better maintenance or improved training to support better sense-making. This kind of thinking leads us to consider the design of the system and its resilience against certain types of error. In well-designed systems, there are barriers to stop human error from 'leaking' into the rest of the system:

- Fail-safe operation – human error causes the system to fail but nobody gets hurt
- Fail-soft operation – the system fails slowly, gives cues that something is wrong so operators have time to prevent a disaster
- Removal of 'latent' hazards and design deficiencies that prompt mistakes

Figure 2.5 presents a flowchart for categorising human error.

Some questions to consider when using the flowchart include whether the action was intentional, whether the person had the right training, the information operators had, the instructions they had been given, whether the person had done the same

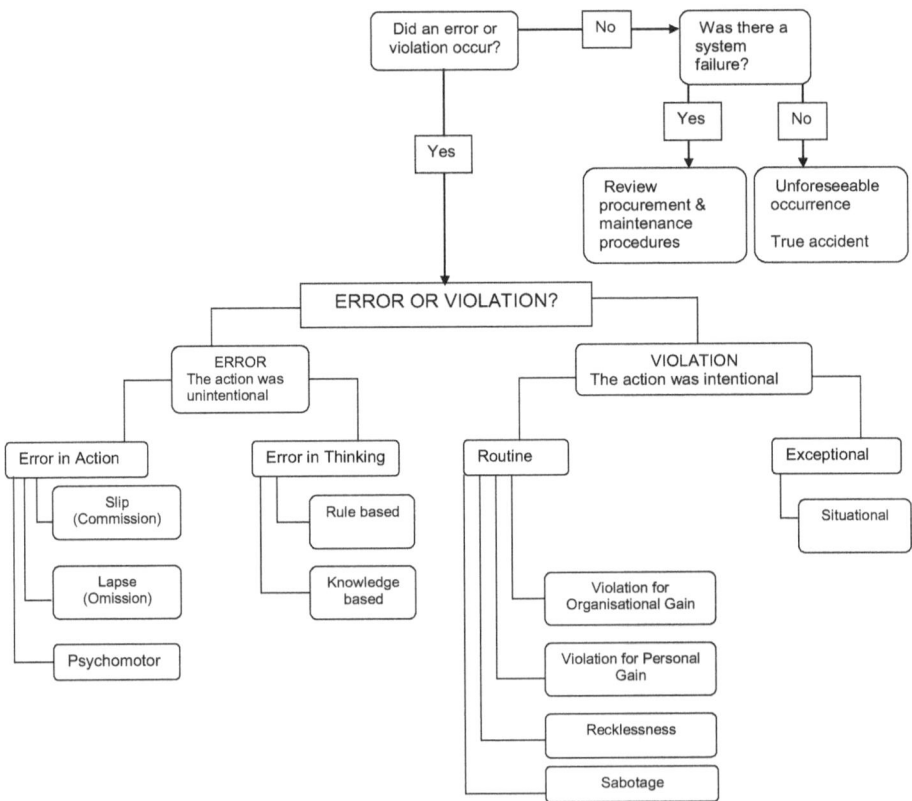

FIGURE 2.5 Flowchart for classifying errors and violations in accident investigations. (From Bridger et al., 2012, Crown Copyright, Contains public sector information licensed under the Open Government Licence v2.0.)

thing before and whether anything unusual was happening at the time. As Dekker (2006) has argued, the identification of human error should be the starting point of an investigation, not the end point. Designers and managers often have the erroneous assumption that human error is rare, whereas most of us are 'accident prone' to a greater or lesser extent as is explained below. Coupled with simplistic notions about cause and effect, these misconceptions may engender undue focus on those involved at the time rather than focussing on the rest of the system where deeper PSFs and causal factors may be found. For example, the probability of a car fatality in the US is twice the probability of a fatality in Norway – this is due to the totality of the system – not that drivers in the US are so much worse.

SITUATION AWARENESS

Endsley (1995) defines three levels of situation awareness:

Level 1, the perception of task relevant elements in the environment
Level 2, the comprehension of their meaning in relation to task goals
Level 3, the prediction of future events

To be 'situationally aware' means having a mental model of the task, 'running' the mental model and updating it as the task progresses. When you imagine yourself carrying out a task, you are 'running' your mental model of that task (imagine yourself reversing your car into a parking space, for example). 'Running' the model requires monitoring one's actions in relation to one's intentions and the state of relevant features of the task environment. Loss of 'situation awareness' is nothing more than a mismatch between one's model of the task status at a particular point and the actual status of the task.

These ideas about situation awareness are useful when systems are being developed. User trials can be conducted using prototype systems and a variety of techniques can be used to assess operator situation awareness. The task can be stopped at critical points and the operator can be asked probe questions to assess his or her understanding of the current state of the system. This can be used to assess the design of the task, identify latent hazards or design faults if critical information is not properly displayed or easy to interpret. Operators can be asked to give a running commentary as they perform the task to assess their level of understanding and identify knowledge gaps for further investigation.

Accidents are often said to have occurred because individuals lacked 'situation awareness' or failed to make the right sense of the situation. It is in the retrospective use of the concept that problems can arise. Terms like 'situation awareness' are only descriptive and they do not describe much. They should not be used as a final explanation for what happened, as they do not explain *why* individuals did what they did or why the accident occurred. What is important is to understand the actual awareness operators had at the time, and why this may have differed from what they should have been aware of, with the benefit of hindsight. We can never be fully aware of everything that is happening around us all the time.

The purpose of conducting safety investigations is to generate safety recommendations to improve the safety of a system. One of the problems that can arise when

FIGURE 2.6 Did accidents happen here because drivers lacked 'situation awareness' or because there is a hazard in the parking bay? For a critique of the concept of situation awareness see Dekker, S.W., 2015. The danger of losing situation awareness. *Cognition, Technology & Work*, 17(2), pp. 159–161. (From Bridger RS, *Introduction to Human Factors and Ergonomics*, 4th Ed, CRC Press, Boca Raton, FL.)

using concepts such as situation awareness to explain safety occurrences is in the kinds of safety recommendations that follow from the explanation. In Figure 2.6, we can see a telegraph pole in a parking bay in a car park.

Protective padding and hazard tape have been placed around the pole as a result of drivers reversing into it. Drivers do not expect to find telegraph poles in the middle of parking bays and it is no surprise that many have reversed into it, prompting (eventually!) the installation of the padding (either by the owner of the car park or the owner of the telegraph pole). It might be concluded that the accidents happened because drivers lacked situation awareness. The safety recommendation that might follow would be to take steps to increase driver situation awareness. However, objects such as telegraph poles do not belong in parking bays – design should help us to be error free – thus the pole should have been removed when parking lot was designed or physically separated from the parking bays. This example illustrates the inadequacy of using the concept of situation awareness and of focusing only on those involved at the time – as we saw at the beginning of this chapter, HF is not about people, but about systems. By focussing on psychological explanations in the immediate vicinity at the time the accident took place, we are limited to a micro-ergonomic analysis that directs safety recommendations towards those involved at the time, whereas we could equally argue that the owners of the telegraph pole and the owners of the car park also lacked situation awareness and were unaware that there was a hazard in the parking bay (until, at least, many drivers had damaged their cars and had complained).

A better explanation is that the accidents happened because of a hazard in the parking bay.

Focussing on the hazard is more likely to lead to effective safety recommendations than focussing on the 'situation awareness' of the drivers or their ability to 'make sense' of the situation they were in.

Auto-detection of Error: How do people become aware that they or their colleagues have made an error? Sellen (1994) described three mechanisms by which people detected their own errors and also identified situations in which the errors were detected by others. The main mechanisms of self-detection are *action-based*, *outcome-based* and *process-based* (in which the constraints in the external world prevent further action). *Action-based detection* involves 'catching oneself in the act' and depends on whether people are monitoring their actions in relation to their intentions (paying attention to what they are doing). It also depends on feedback from the task and therefore on the design of the task and the equipment. An operator who fails to open a pressure relief valve when required may not detect the error if there is no indication that it has not been done or if the feedback is misleading or the system state is not explicitly displayed. Action-based detection can be a form of 'just in time' error detection allowing remedial action. There are several requirements for successful detection of error on the basis of *outcome*. Firstly, the person must have expectations about what the outcome of the behaviour is likely to be; secondly, the effects of the actions must be perceptible; thirdly, the behaviour and the system must be monitored throughout the action sequence; and fourthly, the person must attribute any mismatches between desired and actual outcomes to their own actions and not to the hardware or to extraneous factors. *Process-based detection* occurs when constraints in the world 'short circuit' the sequence of actions. For example, word processors usually prevent users from closing a working file until it has been saved.

In summary, whether people are able to detect the errors they have made depends to a large extent on the design of the task; the design of the objects used to perform the task; the feedback given during the sequence of actions needed to complete the tasks; and their level of knowledge and training – all of which are in accordance with the approach of HF described in Figure 2.1.

VIOLATIONS AND ADAPTATIONS TO UNDERSTAND WORK AS DONE

Whereas errors are unintentional, violations are *intentional* actions or decisions not to follow procedures, rules or instructions. Violations are classed as 'routine' or 'exceptional'. Routine violations are sometimes ubiquitous. Many years ago, trade unions used 'working to rule' as a form of industrial action – by explicitly following every rule to the letter, productivity was lower. Routine violations can happen in highly bureaucratic organisations where new rules are introduced without proper integration, resulting in a plethora of minor rules and procedures. In these cases, violations may even be tolerated by management. This theme has appeared more recently in the safety literature as 'work as imagined' as opposed to 'work as done', which sees violations as simply a way of getting the job done. People may break the rules for organisational gain (so that the organisation benefits) or for personal gain (e.g. to finish work early) or due to recklessness (ignoring the risk) or for intentional sabotage. Exceptional violations are isolated departures from authorised procedures due to unforeseen circumstances, and situational violations are due to situation-specific factors (e.g. excessive time pressure).

The writer was once asked during a safety presentation what to do if employees do not follow standard operating procedures and replied that the first thing to do was to determine what was wrong with the procedures.

Risk Homeostasis Theory (RHT): Wilde (1994) developed RHT during research into road safety. Although RHT is controversial, it may provide insights into the factors that promote rule following and rule breaking. Many, seemingly effective, safety interventions may provide perverse incentives for risk taking that were not considered when the intervention was introduced. RHT starts with the proposition that in any situation where people have control over a process, they adjust their behaviour in relation to a 'target level of risk' which is assessed using feedback from the task. Drivers, for example, may drive faster on wider roads and slow down when it starts to snow, the target level of risk is set by the perceived costs and benefits of engaging in safe or unsafe behaviour. For example, when driving to a business meeting, motorists may drive faster if they left home too late rather than too early. For safety investigations where a violation has occurred, RHT prompts the investigator to search for PSFs that might have prompted the violation (e.g. excessive time pressure, management rewards productivity but not safety). What appear from the outside to be violations may to operators be nothing more than normal behaviour within their 'comfort zone'.

From the perspective of safety management, accidents often happen when people think they are safe. The following needs to be considered: design systems so that operators have an accurate perception of the true risk of accidents as well as the consequences; when designing safety procedures, consider the costs and benefits to operators of following safety procedures; and, if necessary, provide explicit rewards for safety behaviour and not only penalties for unsafe behaviour.

TIME PRESSURE – AN IMPORTANT ISSUE FOR HUMAN FACTORS

Failure to follow standard operating procedures or deliberate violation of such procedures may be a symptom of deeper trouble in the organisation rather than merely negligence or fatigue on the part of the operators. The official report into the grounding of the *Herald of Free Enterprise* found that the failure to close the bow doors of the vessel before it sailed had happened before on other ferries owned by the company. Evidence that Masters were under pressure to leave Zeebrugge 15 minutes early was cited in the official report (Department of Transport 1987):

> The sense of urgency to sail at the earliest possible moment was exemplified by an internal memorandum sent to assistant managers by Mr. D. Shipley, who was the operations manager at Zeebrugge. It is dated 18th August 1986 and the relevant parts of it reads as follows: 'There seems to be a general tendency of satisfaction if the ship has sailed two or three minutes early. Where, a full load is present, then every effort has to be made to sail the ship 15 minutes earlier .... I expect to read from now onwards, especially where FE8 is concerned, that the ship left 15 minutes early ... put pressure on the first officer if you don't think he is moving fast enough. Have your load ready when the vessel is in and marshall your staff and machines to work efficiently. Let's put the record straight, sailing late out of Zeebrugge isn't on. It's 15 minutes early for us'. On the day of the accident, the ferry's departure was 5 minutes late.

ARE SOME PEOPLE 'ACCIDENT PRONE?
CHARACTERISTICS, CAUSES AND MITIGATION

The concept of 'accident proneness' dates from World War 1 when it was found that the statistical distribution of large numbers of accidents in munitions factories did not fit the theoretical distribution (a 'Poisson Distribution') as expected – some people had more accidents than they should have! The concept lost favour in the latter half of the 20th century as attention shifted towards the design of safety systems and equipment in accordance with the focus of HF. More recently, there has been renewed interest in accident proneness and several studies have shown that a measure of cognitive failure in daily life – 'The Cognitive Failures Questionnaire' (CFQ, Broadbent et al. 1998) – can predict safety behaviour. The CFQ has questions about common slips, lapses of attention and memory problems (such as forgetting names) in daily life. Most people report at least few such failures in daily life and some report more than others. Larson et al. (1997) found a link between scores on the CFQ and actual mishaps which concurs with other research that demonstrates that accidents can result from distractibility, mental error and poor selective attention. Wallace and Chen (2005) found that people with high scores on the CFQ were more likely to have accidents due to lower attentiveness. In other words, accident proneness is nothing more than the tendency towards cognitive failure in daily life – we are all accident prone. To date, the evidence indicates that cognitive failure is what psychologists call a 'trait' – a stable characteristic that varies between people.

There is also evidence that accident proneness varies *within* people. Day et al. (2012) found that a link between high scores on the CFQ and having an accident was mediated by psychological strain (caused by exposure to psychosocial stress in the workplace). People under stress are more susceptible to cognitive failures, and these failures then cause accidents.

These findings suggest that accident proneness is both a trait (a stable feature of some individuals) and a state (a condition that can affect all, if under sufficient stress). Since adverse reactions to stress at work increase the propensity for accidents to happen, psychosocial stressors in the workplace can be considered to be PSFs in safety management and should be considered during safety investigations. The concept is more useful when the role of stress is considered, implying that a closer link is needed between health monitoring and safety management. Since it is easier to change working conditions than it is to change people, workplaces can be made safer if they are designed to suit accident prone employees (i.e. if they are designed to be resilient when cognitive failures occur).

Case Study of Poor Human Factors Design –
Grounding of the Cruise Ship *Royal Majesty*

Dekker (2014) gives an interesting analysis of the circumstances leading to the grounding of the cruise ship *Royal Majesty* in June 1995. The ship departed from St. Georges, Bermuda, en route to Boston in the USA. The ship's navigator carried out the required navigational procedures, setting the ship's autopilot to the correct mode and checking the ship's position given by the GPS (Global Positioning System) against the ship's ground-based radar system (Loran-C). The agreement between the two systems was within the required tolerances and the ship departed. Thirty-four hours after departure, the ship ran aground, 17 miles off course.

- June 10, 1995, RM departs St. Georges, Bermuda, en route to Boston.
- Ship's navigator sets autopilot in NAV mode. Checks GPS against ground/radio-based positioning systems (Loran-C). Ground-based system reveals the position about 1 mile southeast of GPS position.
- 30 minutes after departure, the cable from the GPS receiver to the antenna had come loose and the GPS had defaulted to dead reckoning mode (navigating by estimating the distance and direction travelled rather than by using landmarks or astronomical observations in the absence of the GPS).
- 34 hours after departure, ship runs aground, 17 miles off course.

The official report into the grounding contains an extensive analysis of the incident itself: the actions of the captain and the officers on the bridge in the hours prior to the grounding and the equipment and systems in use at the time. At first sight, it appears remarkable that the officers were unaware that the navigation system had defaulted to dead reckoning mode and was not connected to the GPS. They also failed to understand the significance of a red light sighted by a lookout and a call from a Portuguese fishing vessel to warn the ship that it was in shallow water. At some points in the report, there is a focus on what the officers *should* have done, for example, 'Had the officers compared position information from the GPS and the Loran-C they should not have missed the discrepant co-ordinates' and, on approaching land, 'The second officer's response to these sightings should have been deliberate and straightforward. He should have been concerned as soon as the buoy was not sighted and when the lookouts sighted the red light…'. Statements such as these are easy to make with the benefit of hindsight because accident investigators know what happened next. Hindsight bias shifts the focus of an investigation onto what the those involved *should* have done. Focussing on those who were closest in time leads to *hindsight bias* – safety initiatives centred on redrafting of instructions, rules and training procedures.

However, the report (in accordance with the scissors analogy in Figure 2.1) also identified a number of PSFs that help us to understand *why* everything seemed to be in order:

- In those days, satellite reception was poor and the GPS often went 'offline' for short periods. This was not necessarily a cause for concern
- There was no clear indication that the system had defaulted to dead reckoning mode
- The GPS display indicated that the ship was on course
- When the GPS did go offline, the indications that it had done so were weak (a short 'beep' may have been heard and the letters 'DR' appeared at the bottom of a small screen)
- Even if they had heard the 'beep' it was nothing to worry about

- Officers were told not to use the Loran-c (RADAR) when at sea but to rely on the ECDIS (GPS-based) system
- The two systems had been decoupled as a matter of policy, making the system brittle and depriving the officers of important cues.

As is illustrated by the example in Figure 2.3, people are often insensitive to counterfactual evidence when everything seems OK, therefore the salience of the red light may have been lost.

The above is a short summary of selected parts of the official report (National Transportation Board 1997). As was stated earlier, the purpose of HF investigations is to make safety recommendations, and in this respect the report is fairly comprehensive, targeting recommendations at different levels to a number of different authorities:

- To the US coastguard:
- To the Cruise Line
- To the National Marine Electronics Association
- To the developers of integrated bridge systems
- To the International Electrotechnical Commission
- To the manufacturers of ship positioning systems
- To the international council of cruise lines
- To the international chamber of shipping and the international association of independent tanker owners

This example illustrates how the inclusion of HF as an important theme in the investigation increases our understanding of why the incident happened and generates a rich set of safety recommendations to avoid future incidents of this kind.

MAKING SENSE OF SAFETY OCCURRENCES – THE CONTRIBUTION OF HUMAN FACTORS

As stated at the beginning of this chapter, the purpose of safety investigations is to make safety recommendations. HF contributes to this process firstly by analysing the events that took place during a thin slice of time in the immediate vicinity to identify errors or violations (Figure 2.5) in relation to any PSFs (Figure 2.2). A deeper investigation then follows to understand the systemic factors that created the situation in the first place – a macro-ergonomic analysis of the system (Figure 2.7).

SOME KEY CONSIDERATIONS FOR HUMAN FACTORS IN ACCIDENT INVESTIGATIONS

The author has acted as a member of an investigation panel, as a Human Factors Advisor to Service Inquiries in the Royal Navy and as an advisor to Safety Investigation Teams in the UK National Health Service. Based on these experiences,

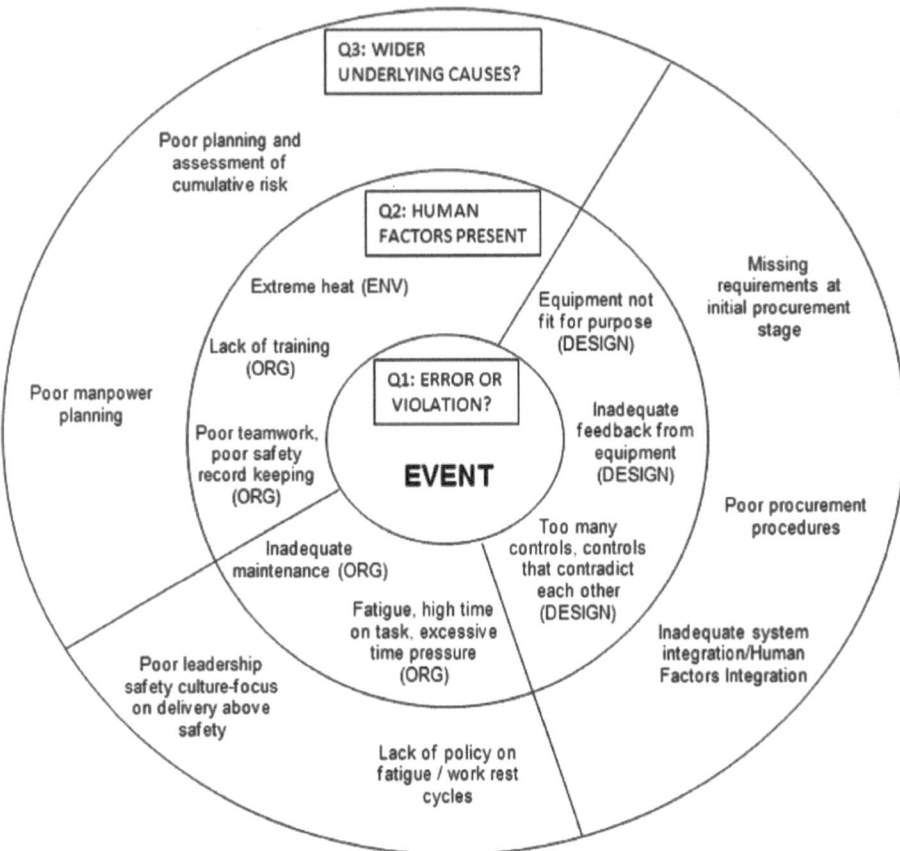

FIGURE 2.7   Overview of a macro-ergonomic analysis.

some key guidelines for successful integration of HF into accident investigation are given below:

- Organisations should ensure that a least one member of the team has HF training or that the team is supported by a suitably qualified and experienced (SQEP) professional in HF. Normally, this means that the individual will be registered with the national HF/Ergonomics Society at a suitable level (is a Chartered Member in the UK).
- Local investigation at site:
  - How did the involved actors understand and make sense of the situation? (Is there any support for resilience or auto-detection of errors?)
  - Were there risk factors for human error in the workspace at the time?
  - Is there any evidence for fatigue, long hours or other organisational, environmental factors that might increase the propensity for human error or violations?

- The investigation should focus not only on immediate events at a local level (site visits, interviews with operators, line managers and so on) but on a range of wider issues:
  - Visit other sites where the same or similar operations are conducted to identify any generic risks and to determine whether any safety recommendations might apply more widely
  - Determine whether similar incidents have happened before, how where they investigated and what was found. How did the organisation respond? Any reports into previous accidents should be reviewed
  - Gain an understanding of the organisation's safety management system, whether HF plays a role and review the Hazard Log (if available)
  - Review any policies and procedures that govern the operations in the accident.
- HF Integration: One way to understand how hazards and risks get into systems is by reviewing the processes the organisation uses to design systems and to manage safety throughout the system life cycle. HF is now integrated into procurement in some organisations via formal policy (e.g. Ministry of Defence 2015). Some features of successful integration include:
  - A HF integration plan was developed at the early stages. Risks were identified early and designed out of the system
  - The requirements for safe operation were specified early on and tests were undertaken to ensure that they were met before the system was accepted into operation
  - Tests involved real users and operators who were consulted early on
  - Lessons from accidents in previous systems were identified and learned
  - Early involvement of Human Factors professionals (SQEP) is mandatory.

The concepts described in this chapter can be usefully applied to the investigation of safety occurrences to provide answers to the questions above. These answers will provide the evidence to support the drafting of safety recommendations, enabling organisations to improve safety by learning from safety occurrences. As Kilskar et al. (2019) have pointed out:

> Sensemaking has often been limited to an organizational context, seldom discussing issues such as system design … it is pointed out that safety science seems to have drifted from the engineering and design side of system safety toward organizational and social sciences or refinement of probabilistic models; thus, there is a need to focus more on design and design principles to be able to diagnose hazardous states in operations

REFERENCES

Bridger, R.S. 2015. Human Factors in Accident Investigation and Safety Management. *Conference on Human Factors in Control: Accident Prevention: Learning and Changing from Investigations*. Britannia Hotel, Trondheim, Norway, 14–15 October 2015.

Bridger, R.S. 2018. *Introduction to Human Factors and Ergonomics*. Fourth ed. CRC Press, Boca Raton, FL.

Bridger, R.S., Johnsen, S., Brasher, K. 2013. Psychometric properties of the cognitive failures questionnaire. *Ergonomics*, 56:1515–1524.

Broadbent, D.E., Cooper, P.F., Fitzgerald, P., Perkes, K.R. 1998. The cognitive failures questionnaire and its correlates. *British Journal of Clinical Psychology*, 21:1–16.

Chartered Institute of Ergonomics and Human Factors. 2020. *White Paper on 'Learning from Incidents'*, Wootton Park, Warwickshire, UK.

Day, A.J., Brasher, K., Bridger, R.S. 2012. Accident proneness re-visited: the role of job strain. *Accident Analysis and Prevention*, 12:532–535.

Def Stan 00-56-Safety Management Requirements for Defence Systems, https://bainessimmons.com/aviation-training/training-courses/tr105-def-stan-00-56-safety-management-requirements-for-defence-systems.

Dekker, S. 2004. *Ten Questions about Human Error: A New View of Human Factors and System Safety*. CRC Press, Boca Raton, FL.

Dekker, S. 2006. *The Field Guide to Understanding Human Error*. Ashgate Publishing, Hampshire, UK.

Dekker, S. 2014. *Safety Differently: Human Factors for a New Era*. CRC Press, Boca Raton, Fl.

Dekker, S. 2015. The danger of losing situation awareness. *Cognition, Technology and Work*, 17(2):159–161.

Department of Transport. 1987. MV Herald of free enterprise Report of Court No. 8074. Formal Investigation.

Drupsteen, L., Guldenmund, F.W. 2014. What is learning? A review of the safety literature to define learning from incidents, accidents and disasters. *Journal of Contingencies and Crisis Management*, 22(2):81–96.

Endsley, M.R. 1995. Towards a theory of situation awareness in dynamic systems. *Human Factors*, 37:32–64.

Gould, K.S., Knappen Røed, B., Koefoed, V.F., Bridger, R.S., Moen, B.E. 2006. Performance-shaping factors associated with navigation accidents in the Royal Norwegian Navy. *Military Psychology*, 18(Suppl):S111–S129.

Kilskar, S.S., Danielsen, B.E., Johnsen, S.O. 2019. Sensemaking in critical situations and in relation to resilience—a review. *ASCE-ASME Journal of Risk and Uncertainty in Engineering Systems, Part B: Mechanical Engineering.* https://doi.org/10.1115/1.4044789

Larson, G.E., Alderton, D.L., Neideffer, M., Underhill, E. 1997. Further evidence on dimensionality and correlates of the cognitive failures questionnaire. *British Journal of Psychology*, 88(1):29–38.

Leveson, N.G., 2016. Engineering a safer world: Systems thinking applied to safety (p. 560). MIT Press, Cambridge, MA.

Ministry of Defence. 2015. JSP 912. Human factors integration in defence systems. https://assets.publishing.service.gov.uk/government/uploads/system/uploads/attachment_data/file/483176/20150717-JSP_912_Part1_DRU_version_Final-U.pdf.

Mosenkis, R. 1994. Human Factors in Design. In: *Medical Devise, International Perspectives on Health and Safety*, C.W.D. van Gruting, ed. Elsevier, New York, pp 41–51.

National Transportation Board. 1997. Marine Accident Report Grounding of the Panamanian Passenger Ship Royal Majesty on Rose and Crown Shoal near Nantucket, Massachusetts, June 10, 1995. PB97-916401 NTSB/MAR-97/0l, Washington DC 20594.

Norman, D. 2013. *The Design of Everyday Things*. Revised and Expanded Edition (2013 ed.).MIT Press, Cambridge, MA.

Sellen, A.J. 1994. Detection of everyday errors. *Applied Psychology: An International Review*, 43:475–498.

Wallace, J.C., Chen, G. 2005. Development and validation of a work specific measure of cognitive failure: Implications for organisational safety. *Journal of Occupational and Organisational Psychology*, 78:615–632.

Wilde, G.J.S. 1994. *Target Risk*. PDE Publications, Toronto, Canada.

3 What Makes a Task Safety Critical?

A Barrier-Based and Easy-to-Use Roadmap for Determining Task Criticality

S. Øie
DNV

CONTENTS

INTRODUCTION

Several approaches to risk and barrier management integrate the contribution of human performance and factors by addressing safety critical tasks. As demonstrated by investigations of accidents, such as the Macondo well blowout, human interaction with modern safety critical systems can be highly complex. Consequently, risk management practitioners are commonly faced with the challenge of pinpointing which tasks should be devoted the most attention and resources. As such it is often necessary to determine the level of task criticality to decide on the need for more in-depth analysis and follow-up by use of improvements in design, procedures and training. When dealing with a large number of tasks, this process needs to be efficient enough to not be overly time consuming and costly. At the same time, it needs to be accurate enough so that critical tasks are not incorrectly categorized (and prioritized). This chapter discusses what characteristics defines a task as safety critical and suggests an easy-to-use roadmap for screening and identifying safety critical tasks based on their level of criticality.

BACKGROUND

Despite continuous work and advancements in risk management, major accidents keep occurring, with the Macondo blowout's 10-year anniversary being a chilling reminder (U.S. Chemical Safety and Hazard Investigation Board, 2016). While the systemic causation of major accidents will be forever mind-boggling, disasters like the Macondo blowout inevitably happen in the wake of increased complexity introduced by concurrent technological and operational developments. For the petroleum industry, which is in this chapter's focus, wells are being drilled longer in deeper waters and in harsher environments.

To cope with the continuously changing risk picture, both the regulatory bodies and the industry actors develop their risk management frameworks, models and requirements to stay within acceptable limits of what is considered safe. One of these developments consists of more systematic considerations of how human performance contributes to major accident risk. For example, the United Kingdom's (UK) Health and Safety Executive promotes analysis and management of human failures when performing safety critical tasks (Health and Safety Executive, 2016). Similarly, the Norwegian Petroleum Safety Authorities (PSA) requires operational barrier elements, a sub-type of safety critical tasks, to be managed according to the same requirements as technical barrier elements (PSA, 2017). Furthermore, the use of human reliability assessments to study safety critical tasks as an integrated part of quantitative risk analysis (QRA) was made increasingly feasible with the launch of the Petro-HRA method in 2017 (Bye et al., 2017). Petro-HRA is the first HRA method specifically tailored to meet the needs of human error quantification performed as part of QRAs in the petroleum industry. It was developed based on SPAR-H method (Gertman et al., 2005), an HRA technique also promoted by NOPSEMA, the Australian regulator of petroleum activities (2020). Several publications (e.g. Bridges, 2011; Myers, 2013) also indicate that HRA is becoming increasingly used as part of Layers of Protection Analysis (LOPA), a risk analysis technique commonly used under the Functional Safety regime (Center for Chemical Process Safety, 2015).

A challenge frequently encountered by practitioners working with analysis and management of safety critical tasks is how to identify and select which tasks to devote the most attention and resources. Such devotion commonly consists of more in-depth analysis of factors influencing human performance (including errors) or development and use of procedures, training or workplace design. Often faced with limitations in both money and time, how to prioritize the correct tasks therefore becomes an important part of the work. Several pitfalls may present itself as part of this process. Due to the inherent complexity and often large scale of the systems being addressed, the list of what can be considered safety critical tasks may grow to become excessively long. This is the most apparent pitfall and implies a possibility that more critical tasks receive too little attention, relatively, compared to less critical tasks. Alternatively, tasks may be incorrectly labelled as having low criticality and consequently screened out during the selection process. This could happen as a result of insufficient knowledge about the task. A compensating measure is to obtain the knowledge necessary to make informed decisions. This can easily become a time-consuming task and turn into a case of *overcompensation*, unless one has the right tools for the job.

As such, determining task criticality easily presents itself as an ungrateful piece of work where spending enough time, but not too much, must carefully be balanced to achieve the desired outcome. The aim of this chapter is therefore to expand the risk management practitioners' toolbox by providing an easy-to-use roadmap for determining task criticality.

MANAGEMENT OF OPERATIONAL BARRIER ELEMENTS

As explained above, identification and selection of safety critical tasks may be relevant for work both as part of safety studies and risk management. The lessons learned, described in this chapter, mainly stem from working with implementation of systems for managing *operational barrier elements* on various O&G installations on the Norwegian Continental Shelf (NCS), in both the design and the operational phase.

Operational barrier elements are part of the framework for barrier management promoted by the Norwegian PSA both as part of their regulations and industry guidelines (PSA, 2017). Other industry actors have also published guidance on barrier management, some with a more practical orientation than what is currently communicated by the PSA. One such guideline is made available by the Norwegian Shipowners' Association (NSA) (Øie et al., 2014), from which most of the definitions in this chapter are taken from.

An operational barrier element is a type of safety critical task and can be defined as a task performed by an operator or team of operators, which realizes one or several barrier functions. Here "realize" refers to performing the function when it is required, which could be continuously or on demand, depending on the barrier's performance requirements. The barrier function is in most cases realized through interactions with technical barrier elements. These are engineered systems, structures or other design features which realize one or several barrier functions. The barrier function itself is the role or purpose of the barrier in either preventing a hazard from being released, controlling the accident escalation pathway, or mitigating undesired consequences in case an event should occur (Figure 3.1).

Examples of operational barrier elements are provided in Table 3.1. A separate column titled "Systems" has been added to aid the reader's understanding by indicating which technology is involved.

On the NCS operational barrier elements fall under the same regulations as technical barrier elements – §5 Barriers in the Management Regulations (PSA, 2017). In brief, §5 requires that barriers shall be implemented to manage risks emerging from the major accident hazards (MAHs) associated with the installation's operations and environment. Requirements for barrier performance (e.g. functionality, integrity, robustness) shall be defined based on the risk picture, and a plan for how to monitor and maintain such performance shall be established and adhered to. The implementation of barrier functions and setting of performance standards (i.e. strictness of requirements) shall take into consideration the specific context of each area on the

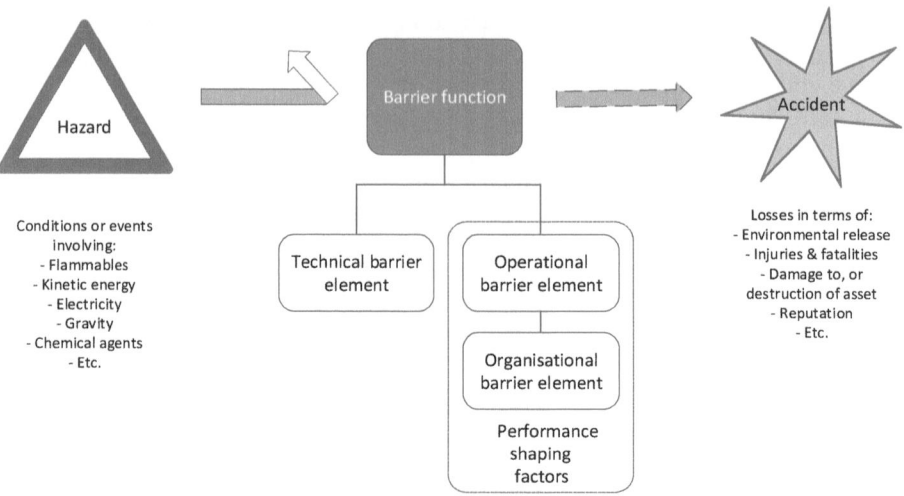

FIGURE 3.1 Barrier functions and elements preventing hazards from causing accidents.

TABLE 3.1

Examples of Major Accident Hazards and Associated Barrier Functions and Operational Barrier Elements (Simplified)

Major Accident Hazard	Barrier Function	Operational Barrier Element	Systems
Loss of position	Prevent damage to wellhead	Perform emergency disconnect from well	Marine/drilling
Leak in firewater ring main	Prevent loss of stability/capsizing	Shutdown firewater pumps	Marine/utility
H_2 gas in battery room	Prevent ignition of H_2 gas	Electrically isolate the battery room	Utility
Well kick	Prevent well release	Shut in well by closing BOP	Drilling
Hydrocarbons above subsea blowout preventer	Prevent ignition of hydrocarbons	Activate the diverter system to divert hydrocarbons away from rig	Drilling
Hydrocarbon leak in process area	Reduce ignition probability/ explosion pressures	Perform emergency blowdown	Production
Hydrocarbon leak in risers or pipelines	Reduce/limit hydrocarbon leak	Shut down oil export pumps and close riser emergency shutdown valve	Production

installation, e.g. by considering the presence of hazards and conditions which may deteriorate barrier performance. This forms the installations safety (barrier) strategy; the results from a process that, based on the risk picture, describes what barrier functions and barrier elements shall be (or have been) implemented in order to reduce risk (PSA, 2013). For operational barrier elements this translates into ensuring that the personnel performing them, referred to as *organizational barrier elements*, are provided with the competence, procedures and other workplace aids required to execute such actions in a reliable manner. In a design phase, focus will be on ensuring satisfactory level of quality on more technical factors influencing human performance, such as alarm system, use of automation and various human–machine interfaces.

A PROCESS FOR MAPPING AND ASSESSMENT OF OPERATIONAL BARRIER ELEMENTS

The approach to implement management of operational barrier elements, in which determination of task criticality was required, consists of five main steps. This stepwise process was partly based on the guidance provided by the Health and Safety Executive (1999) and the Energy Institute (EI) (2011). However, significant modifications were made to accommodate the framework, definitions and practices of barrier management, in particular those suggested by the NSA (Øie et al., 2014) and PSA (2017). The approach is briefly summarized to provide the reader with some additional understanding of the context (i.e. process) in which ranking of task criticality was performed.

Step 1. Identification. The first step is to identify safety critical task which could be categorized as operational barrier elements. This consists of reviewing various documents such as reports about risk analysis and safety studies, system and operational philosophies, procedures and manuals, incident and maintenance records, and more. Feedback about experience from operational personnel is another valuable source of information.

Step 2. Screening. The second step is to screen which operational barrier elements should be subject for further evaluations and follow-up based on a ranking of criticality against a set of pre-defined criteria. Tasks are categorized as either being of high, medium or low criticality. This determines how the safety critical tasks are followed up as part of engineering and operations, as descried in Table 3.2.

Steps 1 and 2 make up the main topic of this chapter.

Step 3. Analysis. The third step is to perform task and human error analysis to understand the actions involved, how they can fail, as well as potential influence of performance shaping factors. Task analysis was performed on all tasks ranked as having medium and high criticality while detailed human failures was only studied for tasks ranked as having high criticality.

TABLE 3.2
Criticality Levels Used as Criteria to Decide on Follow-Up Actions

Criticality	Detailed Analysis?	Coarse Task Analysis?	Implemented?
High	Yes	Covered by detailed	Yes
Medium	No	Yes	Yes
Low	No	No	No

Step 4. Evaluation. The fourth step is to evaluate task characteristics (complexity, time available, communication, etc.) and failures to optimize human performance (reliability) through improvement in design features (e.g. HMI and alarm systems) as well as to generate requirements for how the operational barrier element shall be executed, when and by who.

Step 5. Implementation. The fifth and final step consists of implementing the findings from the evaluation (Step 4) into relevant design specifications, operational and emergency procedures, other governing documents, and training or competence programs. An important part of documenting the operational barrier elements and associated performance requirements was to include them as part of the installations safety (barrier) strategy.

A more detailed description of an operational barrier element mapping and assessment case study can be read in a chapter written by Ludvigsen et al. (2018).

GUIDANCE (MISSING) ON HOW TO IDENTIFY OPERATIONAL BARRIER ELEMENTS

Developing the process for mapping and assessing operational barrier elements started with reviewing available methodologies for *Step 1: Identification.* The guidance for safety critical task analysis issued by Health and Safety Executive UK and EI both contain tools and examples for how to identify safety critical tasks and rank their criticality. These were therefore reviewed as a starting point towards establishing a suitable approach.

Health and Safety Executive UK's guidance recommendation for identification of safety critical tasks is to have people "brainstorm" around certain topics such as equipment, phases of operation, task types and job descriptions. However, no specific methodological approach is described. Instead, a set of "generic critical task inventories" are provided to use as a starting point. As it turned out, the listed tasks mainly reflect high-level operational activities, such as those related to separation, oil export, gas dehydration and gas compression for production facilities, and drilling, pumping, reeling and hoisting for rig operations. Due to being described at a relatively high level of detail, checks revealed that subsequent determination of task criticality levels proved difficult without closer examination, which in turn would defeat the purpose of safety critical task identification and screening. Subsets of operations required for handling abnormalities and emergencies, which are closer to the definition of operational barrier elements, were covered to a less extent. As such, neither the task inventory nor the safety critical task identification approach suggested by Health and Safety Executive UK could be applied directly for the purpose of identifying operational barrier elements.

To further understand the specific needs behind how to identify operational barrier elements, it can be useful to put the definition into a broader context. The PSA distinguishes between risk reduction through use of safe and robust solutions, and risk reduction by implementation of barriers (PSA, 2017, see Figure 3.2).

"Robust solutions" refer to those which prevent the likelihood of hazardous events being initiated in the first place, such as use of inherent safe design principles by removing hazards instead of controlling them. It may however also refer to operations

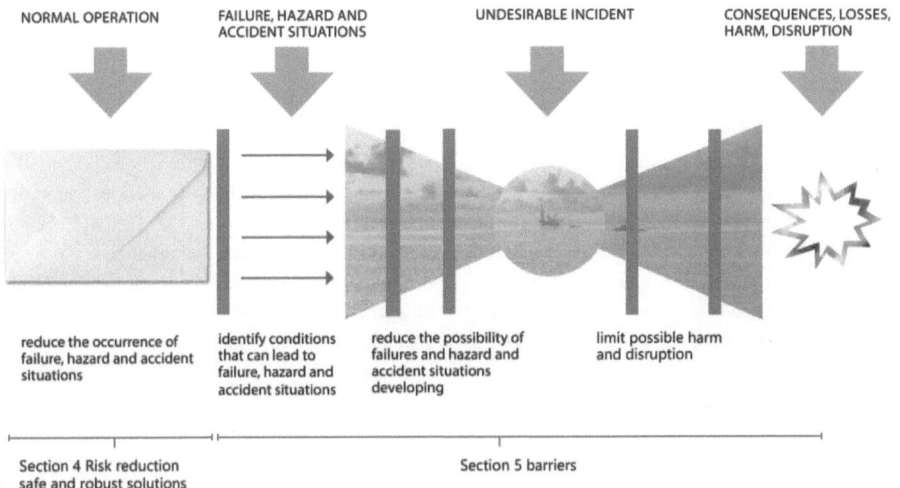

NORMAL OPERATION   FAILURE, HAZARD AND ACCIDENT SITUATIONS   UNDESIRABLE INCIDENT   CONSEQUENCES, LOSSES, HARM, DISRUPTION

reduce the occurrence of failure, hazard and accident situations   identify conditions that can lead to failure, hazard and accident situations   reduce the possibility of failures and hazard and accident situations developing   limit possible harm and disruption

Section 4 Risk reduction safe and robust solutions   Section 5 barriers

FIGURE 3.2 The distinction between "barriers" and "robust solutions" (PSA, 2017).

such as how criteria for operational envelopes are defined, e.g. in terms of which weather and ocean conditions are allowed under.

Barriers refer to functions which can identify triggering events and prevent them from developing into an undesirable incident or which can mitigate possible losses in case such incidents should occur. Safety critical tasks characterized as operational barrier elements arguably first come into play as part of functions for monitoring hazardous conditions and detecting failures with potential for escalation. Arguably they are not to be considered a part of ensuring as "robust solution". This implies that operational barrier elements are partly "pre-selected" as safety critical task through how they are defined as being part of a barrier function implemented for MAHs.

The most noticeable challenges when identifying operational barrier elements is to decide on where to draw the line between safety critical tasks being a part of the robust solution and safety critical tasks being a part of barrier functions (i.e. operational barrier elements). As an aid in making this distinction, the categorization of safety critical tasks into Type A, B and C actions can be useful (Øie et al., 2014). As shown in Table 3.3, operational barrier elements are either categorized as Type C and, in some cases, Type B actions (please note that the terms "failures" and "incidents" here refer to those related to MAHs and not to occupational accidents).

Type A actions falls outside the definition of what is considered an operational barrier element because it does not involve continuous or *on demand* monitoring and control of barrier functions. For Type B actions, it may not always be clear cut whether the safety critical task represents an operational barrier element or not. Examples can be performing a crane lifting operation with the potential of dropped objects onto critical structures or maintaining containment during maintenance on equipment containing pressurized hydrocarbons. In such cases, precautionary steps necessary to prevent a MAH incident from being triggered (Type A) often coincides with actions taken to observe and deal with initiating events (Type C). While the former could be

TABLE 3.3
Type A, B and C Actions

Type	Description	Operational Barrier Element?
Type A	Actions where operator error can introduce or fail to reveal a latent failure.	No. Type A actions are associated with activities potentially influencing barrier integrity, such as installation, operation, inspection, testing and maintenance.
Type B	Actions where operator error contributes directly to initiation of an incident.	Maybe. Type B actions rely on safe and reliable performance to maintain control of the hazard and prevent directly triggering undesired events.
Type C	Actions where operator error allows an already initiated incident to escalate.	Yes. Type C actions are often associated with responses to release of a hazard to control and mitigate escalation of accident scenarios.

Source: Adopted from Øie et al. (2014)

argued to represent a "robust solution", and not an operational barrier element part of a barrier function, the distinction from the latter becomes more theoretical than practical – they can both be potential safety critical tasks, they happen close in time and are performed by the same personnel. In principle, if a single human failure has the potential to trigger a MAH, preventing this from happening by performing such tasks *correctly* could be claimed to represent an operational barrier element. When identifying such tasks, a decision must therefore be made whether the task should be considered an operational barrier element or not, based on the initially perceived criticality. On the NCS or companies operating under similar regulatory regimes, the practical implications from such a decision will be whether the safety critical task is followed up as part of the systems implemented to manage safety barriers.

Both the categorization of safety critical tasks as either Type A, B or C actions and PSAs concept of "robust solutions" vs. "barriers" suggest that operational barrier elements should be identified through which MAHs they are used as barriers for (a list of typical MAHs can be found in Annex C of the NORSOK Z-013 standard). Remembering the definitions of barrier functions and elements, the identification process turns into more of a top-down than bottom-up (as suggested by the Health and Safety Executive UK guidance):

Major accident hazards → barrier functions → technical barrier elements → operational barrier elements (Types B and C)

A practical example could be:

Hydrocarbons → prevent leak → piping and valves → maintain containment during maintenance on pressurized equipment

This approach is somewhat more in line with the recommendations provided by the guideline published by the EI (2011). The first step in their guideline is to identify "main site hazards" (i.e. MAHs). For their second step, identification of safety critical tasks, they recommend looking into documentation which specifically addresses those hazards. Safety critical tasks (and operational barrier elements) can be identified as either causes or safeguards in safety studies such as hazard identification (HAZID)/hazard and operability (HAZOP), LOPA and QRA and also as part of other documents such as operating and emergency procedures, safety strategies and philosophies, as well as technical specifications and requirements.

It should be noted, however, that EIs guideline does not apply the term "barrier function" or the Type A, B and C categorization directly as part of their safety critical task analysis method. Instead, this is partly addressed implicitly through how they define safety critical tasks as tasks where human factors (HF) can contribute to MAHs in positive or negative ways, including:

- Initiating events;
- Prevention and detection;
- Control and mitigation, and;
- Emergency response.

However, despite this definition, the examples provided in the guideline suggest that safety critical tasks are analysed bottom-up according to their operational goals (e.g. "unit start-ups") instead of a top-down approach based on the safety critical tasks (i.e. operational barrier element) role in performing barrier functions (e.g. maintain containment during work on equipment with pressurized hydrocarbons). As such, the approach developed for identification of operational barrier elements was inspired by EIs guidance, but still adopted with the modifications necessary to better align with the framework and definitions of barrier management.

WHAT MAKES A TASK SAFETY CRITICAL?

The next step of the process (Step 2) – and the core topic of this chapter – involves screening the operational barrier elements to determine their level of criticality. Guidance provided by Health and Safety Executive UK and EI was again reviewed for inspiration regarding method development. The former contains a comprehensive approach which involves rating the criticality of offshore production tasks or well operations based on scoring the answers to a set of pre-defined questions. Task criticality is ranked as high, medium/high, medium, medium/low or low based on which intervals the diagnostic score falls within. The same approach was used by the developers to risk rank the generic task inventories published as part of the guidance note.

As it turned out, when attempts were made to apply the method for rating the criticality of operational barrier elements, several challenges emerged (see Table 3.4).

While the questions used to diagnose criticality appear relevant, to some degree, the way they are phrased makes them more suitable for normal production activities and less for operational barrier elements. This also reflects the "generic tasks" listed in the guidance note for which they have been used to rank criticality. Another challenge was that the scoring system was not calibrated to differentiate between operational barrier elements with higher and lower criticality. This was apparent by how

TABLE 3.4

Questions Used to Evaluate Task Criticality in the Health and Safety Executive Guideline (Health and Safety Executive, 1999)

Questions for Diagnosing Task Criticality	Relevance for Operational Barrier Element
1. How hazardous is the system involved?	Also relevant for operational barrier element. The potential severity of the major accident hazard for which the operational barrier element is required would help determine its criticality.
2. To what extent are ignition sources introduced into/during the task?	More relevant for production tasks. Ignition source control is an operational barrier element, and very few (if any) technical barrier elements which operational barrier elements interact with would be designed in such a way that ignition sources are introduced.
3. To what extent does the task involve changes to the operating configuration?	More relevant for production tasks. Useful for checking whether an operation could, for example, cause a leak of hydrocarbons.
4. To what extent could incorrect performance of the task cause damage?	Partly relevant for operational barrier element. However, the phrasing should concern whether incorrect performance could cause/fail to prevent escalation of an event.
5. To what extent does the task involve defeating protection devices?	More relevant for production tasks. When the task itself is a barrier, it is not useful to ask whether it involves defeating other barriers.

the operational barrier elements listed in the guidance note's generic task inventory, such as "response to emergency shutdown", were all ranked as highly critical event without being scored. Lastly, the scoring system requires that all five questions are answered for all tasks (i.e. one score between 1 and 3 per question). In case of dealing with a large number of operational barrier elements, for which not all the questions may be relevant, this can represent a rather daunting and time-consuming task.

Not being able to fully utilize the tools provided by Health and Safety Executive UK, an attempt was made to apply the methods suggested by EI's guidance for safety critical task analysis. In addition to referring to Health and Safety Executive UK's guidance note, EI suggests some more simplified tools, e.g. a three-by-three decision matrix with "consequences of human failure" on one axis and "level of human involvement" on the other. With only two generic factors to consider and a decision matrix as an intuitive and easy-to-use tool, the simplicity of EI's approach offers some benefits in terms of lowering the user threshold. However, with operational barrier elements already being preselected through their association with MAHs and barrier functions, it turned out to be difficult to differentiate which tasks should be assigned with high and medium priority. Generally, most operational barrier elements turned out as having high priority, few with medium and none with low. A more sophisticated matrix is also provided as part of the guidance note's annexes. This does not consider level of "human involvement" but instead includes two other factors in addition to "consequence", namely, task complexity and task frequency. This, however, is not accompanied by any definitions of the various levels, and criticality is primarily indicated in terms of how strict the use of procedures should be during operations. As such, this matrix did not prove to be a viable solution.

Running out of options, the final solution was to develop a method specifically suitable for operational barrier elements. As shown in Table 3.2, the purpose of the criticality ranking was to support decisions about whether to perform detailed or coarse analysis of the operational barrier element (Steps 3 and 4) and whether to implement the operational barrier element as part of the system for barrier management (Step 5). Experiences made from attempting to apply the methods published by Health and Safety Executive UK and EI suggest that the criticality ranking should include certain properties. It should be

- as accurate as possible to determine scalable (high, medium and low) task criticality, but without having to perform much, or if any, analysis and
- generic in the sense that it allowed ranking of operational barrier elements associated with a wide variety of systems and operations (marine, production, drilling, etc.).

Although neither of the available guidelines included a directly applicable method for assessing criticality, together they offer valuable insights about what makes a task safety critical. These insights were combined with the context provided by the framework and definitions of barrier management to suggest a set of characteristics of what determines task criticality.

Major accident hazard severity. The first question to be asked is if the operational barrier element is associated with a barrier function implemented to manage risks associated with MAHs. Establishing this link was also suggested by the EI guideline (2011) as part of the safety critical task identification, but it was not made part of the tools recommended for rating criticality. What constitutes a major accident may vary from company to company and/or from regulator to regulator. Most companies do, however, have this defined as part of their risk management philosophy, e.g. as risk acceptance criteria operationalized by use of risk matrices.

A MAH with potential high level of severity could be one which immediately results in multiple fatalities (e.g. 5) if triggered, such as process fires. A MAH of medium severity could be one which is limited to cause injuries or a lower number of fatalities and only results in major accident losses in case of escalation. An example could be an electrical fire in the utility area of an installation which does not have an immediate effect on any main safety functions and which requires significant escalation to cause losses defined as a major accident. A MAH of low severity only has the potential to cause losses defined as occupational accidents (e.g. injuries or single fatalities) or minor environmental impacts.

Definitions of the "major accident hazard (MAH) severity" levels are described in Table 3.5.

Consequence (from task failure). The next characteristic of a safety critical task is to what degree the consequences from a failure have the potential to contribute to the causation or realization of a major accident. This was also covered by both the EI and

TABLE 3.5

Major Accident Hazard (MAH) Severity Levels

MAH Severity	Levels
High (H)	The hazard involves potential ignition of large HC releases or other incidents which immediately cause a major accident
Medium (M)	The hazard involves incidents which require escalation to cause a major accident
Low (L)	The hazard requires significant escalation to cause a major accident and is therefore considered unlikely

TABLE 3.6

Consequence (from Task Failure) Levels

Consequence	Levels
High (H)	Human failure could alone initiate or fail to prevent/mitigate a major accident
Medium (M)	Human failure could initiate or fail to prevent/mitigate a major accident, but only if other safeguards fail (e.g. automation, equipment)
Low (L)	Human failure should not initiate or fail to prevent/mitigate a major accident

Health and Safety Executive UK guidance, albeit in somewhat different ways. The level definitions presented in Table 3.6 are adopted from EI's guideline, only with some minor adjustments. In case human failure alone has the potential to initiate or fail to prevent/mitigate a major accident, the consequence is high. With some exceptions, such as in drilling and marine operations, most offshore petroleum systems nowadays are, however, well protected and error tolerant, and so trend is that this is increasingly less likely to be the case. Emergency systems, such as those responsible for shutting down the plant in case of abnormalities, are often fully automated and only require the operator to intervene in case of automation failures. Alternatively, there may also be scenarios where the procedure instructs the operator to activate systems even before automation kicks in, as an additional safeguard. In such cases, the consequence from human failure is considered medium. If humans have little interaction with the systems associated with the MAH, or if the system is very well protected, the consequence is low.

Definitions of the "consequence" levels are described in Table 3.6.

Dependency (on human actions). A third characteristic, specific to operational barrier elements, is the degree to which a barrier function relies on human actions to be realized (i.e. performed). This mostly reflects the degree of automation and coincides with the definitions of high, medium and low for "Consequence". The definition was nevertheless kept as a way of creating awareness about the task's role as part of a barrier function. For example, depending on the information available about the task, in some cases it may be easier to determine dependency than consequence, and vice versa. In the method developed for criticality ranking, the dependency and consequence characteristics were therefore combined into one factor, as described later in this chapter.

TABLE 3.7
Dependency (on Human Actions) Levels

Dependency	Levels
High (H)	Barrier function relies on operator actions
Medium (M)	Barrier function relies operator actions in case other safeguards fails (e.g. automation, equipment)
Low (L)	Barrier function relies does not rely on operator actions (no opportunity to intervene)

TABLE 3.8
Unfamiliarity (of Task) Levels

Unfamiliarity	Levels
High (H)	The operational barrier element is novel or non-routine and involves use of new/unfamiliar technology
Medium (M)	The operational barrier element is familiar, but not necessarily routine, and requires specific measures to maintain familiarity
Low (L)	The operational barrier element is familiar, regularly practiced, routine and/or well documented

Definitions of the "Dependency" levels are described in Table 3.7.

(Un)familiarity (of task). One of the tools included in the EI guideline (2011) included "frequency" as a parameter to consider when determining task criticality. While this is a relevant task characteristic, it is argued here that the term can be broadened to also include aspects of *familiarity*. Familiarity can be obtained in various ways: through relevant and available procedures, regular practice through frequent execution of the task (i.e. routine) and by providing various forms of competence activities (drills, table-tops, pre-job meetings, etc.). A highly unfamiliar operational barrier element is characterized by being novel and non-routine, e.g. by having no procedures or training in place, and not being regularly practiced. An operational barrier element of medium familiarity is known to the operators but requires specific measures to maintain familiarity, such as regular drills. In case an operational barrier element is regularly practiced, e.g. as part of normal operations, and is part of well-established routines and procedures, the degree of unfamiliarity is low.

Definitions of the "Unfamiliarity" levels are described in Table 3.8.

Complexity (of task). Task complexity is the final task characteristic included as part of the method for evaluating operational barrier element criticality. The rationale is that the criticality of even the most familiar task can increase if the complexity is high enough. Complexity was also suggested as a factor by EI (2011), however, without providing any specific definition or description. Instead, the definition of complexity in the Petro-HRA guideline was used as an inspiration for defining the ranking levels. A highly complex task that involves a relatively large number of actions which are performed in non-linear sequences (e.g. iterations and parallel)

TABLE 3.9
Complexity (of Task) Levels

Complexity	Levels
High (H)	The operational barrier element involves large and non-linear sequence of actions (trouble shooting, parallel actions, multiple goals, extensive coordination/communication, etc.)
Medium (M)	The operational barrier element is not linear but involves a manageable number of actions so that an overview can be obtained (by most people involved) from the onset
Low (L)	The operational barrier element is straight forward, obvious (i.e. clear cues), involves few steps, and is well supported by HMI/automation

may have multiple goals. Trouble shooting may require knowledge beyond what can be proceduralized, and extensive coordination and communication between several people is part of the task. A medium complex operational barrier element has many of the same characteristics as those with high complexity but to a lesser degree, most often due to a lower number of actions. Operational barrier elements with low complexity are tasks which are highly obvious, e.g. due to a combination of few steps and being supported by well-designed human–machine interfaces, alarm systems and automation.

Definitions of the "complexity" levels are described in Table 3.9.

ROADMAP FOR DETERMINING OPERATIONAL BARRIER ELEMENT CRITICALITY

After having defined the characteristics of task criticality (including levels), the next step was to convert these definitions into a method which solves the challenges experienced with the tools provided by Health and Safety Executive UK and EI. The review of these tools revealed how complex the concept of task criticality is. This, in turn, suggested that developing a quantitative tool with numerical scoring systems may not be feasible or useful, or it would result in a time- and resource-consuming job due to the necessity of complex models. As such, the philosophy was to make a qualitative method which allows for some flexibility and expert judgement. Another aspect of the same philosophy was that it should be possible to determine criticality without necessarily having to consider or document the rationale behind the score for all the factors.

The "roadmap" is presented in Table 3.10. When doing the actual ranking, the available knowledge about the operational barrier element is compared against each factor going from left to right.

As can be read, if both MAH severity and consequence/dependency are both rated High and both combinations of unfamiliarity and complexity consist of one high and one medium (or higher) score, the operational barrier element is considered having high criticality. The criticality is medium as long as either unfamiliarity or complexity is considered low, even if the all the three other factors are ranked as high. If

TABLE 3.10
"Roadmap" Used to Determine Task Criticality

MAH Severity	Consequence/ Dependency	Unfamiliarity	Complexity	Criticality	Analysis	Implemented?
High	High	High	≥Medium	High	Detailed	Yes
High	High	≥Medium	High	High	Detailed	Yes
High	High	≤High	Low	Medium	Coarse	Yes
High	High	Low	≤High	Medium	Coarse	Yes
High	Medium	≤High	≤High	Medium	Coarse	Yes
Medium	High	≤High	≤High	Medium	Coarse	Yes
Medium	Medium	≤High	≤High	Medium	Coarse	Yes
≤High	Low	NA	NA	Low	No	No
Low	≤High	NA	NA	Low	No	No

either MAH severity OR consequence/dependency is high or medium, the overall criticality is also medium, regardless of how unfamiliarity or complexity is scored. Lastly, in cases where either MAH severity OR consequence/dependency is ranked as low, the operational barrier element is automatically also considered to have low criticality, regardless of how other factors are scored. It should also be mentioned that in practice, if the MAH severity is low, the consequence/dependency is also low (if no MAH is present, there are likely no barriers for it, and human failure cannot initiate major accidents).

A keen eye may observe that MAH severity and consequence/dependency are given a higher weighting than unfamiliarity and complexity. The reason is that the two former factors represent more inherent properties of the task (i.e. operational barrier element) than what is the case for the two latter, which are to be considered more like performance influencing factors. In other words, a task cannot be considered critical if there is no MAH or human–system interaction; however, they may still be critical despite being completely familiar and simple. As shown in Table 3.11, this is reflected in the tool by assigning twice as much weight to the levels in MAH severity and consequence/dependency ($H = 3$, $M = 2$, $L = 1$), as for unfamiliarity and complexity ($H = 1.5$, $M = 1$, $L = 0.5$, $NA = 0$). When calibrating the tool, this resulted in the following lower and upper bound intervals for the various criticality levels:

- High criticality: 9–8.5
- Medium criticality: 8–5
- Low criticality: 4–2

As mentioned previously, a numerical scoring system was purposefully not made part of the tool, and so the values were only used for calibrating the combination of factors and levels. Instead, the intention was to use the tool as a look-up table for assigning criticality, without the need for further analysis and documentation. For most practical purposes, it is the level of criticality which is of primary importance,

TABLE 3.11

Scores Used for Calibrating the "Roadmap", Including Upper and Lower Bounds for Per Criticality Level

Levels	MAH Severity and Consequence/Dependency	Unfamiliarity and Complexity	Intervals (Ref. Table 10)	
			Upper Bound	Lower Bound
High	3	1.5	9	8.5
Medium	2	1	8	5
Low	1	0.5	4	2
NA	NA	0		

since this is what determines how it will be followed up as part of analysis or implementation as part of design processes or organizational efforts. Knowing which combinations of factors and levels made the operational barrier element high, medium or low in criticality does not determine what actions are taken further in terms of analysis and implementation efforts. One exception may arise in cases where operational barrier elements are rated as highly critical and where follow-up actions may require substantial costs and/or resources. If this occurs, documenting the rationale behind why the operational barrier element was ranked as highly critical may serve a purpose as basis for informed decision-making. A qualitative analysis would in any case be the most useful.

Experiences from using the tool as part of processes for mapping and assessing operational barrier elements show that an expected distribution in criticality of approximately:

- 10%–20% operational barrier elements with high criticality
- 40%–50% operational barrier elements with medium criticality
- 30%–40% operational barrier elements with low criticality

The distribution depends on what type of technology and operations are being considered, as well as the age of the installations. For example, operational barrier elements related to well control in drilling operations still involve many manual tasks as compared to production facilities with automated fail-safe systems. Similarly, older plants (in general) were designed and constructed at a time when automation was less advanced.

WHAT ABOUT LATENT FAILURES (TYPE A ACTIONS)?

The presented process and roadmap have proved useful for screening operational barrier elements in several large-scale projects, both as part of engineering and during the operational phase. It does however lack a feature which was particularly emphasized in the guidance provided by Health and Safety Executive (1999), namely identifying and screening Type A actions with the potential of introducing latent failures. Although developed with the intention of being applied to operational barrier elements, the roadmap can in principle be used for any safety critical task categorized as either Type B or Type C actions. However, for the purpose of ranking the criticality of Type A actions, the definitions of consequence and dependency (in particular) would have to be significantly altered. As made famous by James Reason (1997), the "Swiss Cheese" model of accident causation explains how most, if not all, major accidents involve latent failures.

A case could be made that safety critical task and barrier management approaches are not necessarily the most suitable concepts for managing risks associated with Type A actions. Such actions are typically plentiful, but at the same time well hidden inside tasks and operations with less than obvious criticality. Most systems nowadays are well defended by use of multiple layers of safeguards, such as interlocks. Spotting critical details from the surface may therefore prove to be difficult. For the same reasons, they would not only be resource demanding to identify and screen but also follow-up in case they would fall under similar requirements as (for example) the PSA expects for operational barrier elements, which can be considered too strict and comprehensive for such tasks. It is also possible that a safety critical task analysis of such tasks would result in diminishing returns. The author has been involved in performing detailed human error analysis of tasks identified as causes of hazardous events logged during a HAZOP study. As it turned out, the causes mainly consisted of Type A actions such as someone having left a valve in wrong position or failed to reinstate an interlock after doing maintenance. These were often single actions and commonly isolated from the overall goal of the task (i.e. low on complexity). A great deal of efforts was therefore spent analysing parts of the overall task which had little or no risk attached to it.

It is argued here that other alternative methods may be more suitable for targeting Type A actions, both in terms of analysis and following up with improvements in design, procedures and training. For example, most of the commonly used safety studies and risk analysis methods can be applied in ways which allow systematic considerations of human error and HF aspects. Techniques such as HAZID, HAZOP, LOPA and failure mode, effect and criticality analysis (FMECA) can all be used to identify not only Type A and B actions as causes but also Type C actions as safeguards (e.g. alarm responses). The same techniques also often include risk rankings which will reveal the associated hazards degree of severity, e.g. by use of risk matrices or similar tools. This can be used to determine task criticality, without having to do a separate screening covering all the safety critical tasks. Instead a criticality ranking tool, such as the presented roadmap, could be used to assess the most critical tasks. Following the argument above about Type A actions commonly being single task steps detached from the overall operation, it would be particularly important to consider the level of complexity. The results from such reviews could then be used to apply basic HF principles for design of HMIs, local control panels, equipment layout, etc., using simple-to-use tools such as checklists.

During operations, some faith must be put into the management systems' and leaders' ability to ensure high quality procedures and training, strong safety cultures and systematic use of maintenance and reliability programs. Not all safety critical tasks should require HF analyses to be properly managed.

CONCLUSIVE SUMMARY

This chapter summarizes the experiences from developing and applying a tool for ranking the criticality of operational barrier elements. Operational barrier elements are safety critical tasks required to realize (i.e. perform) barrier functions implemented to prevent, control or mitigate hazardous conditions and events with potential of causing major accidents. The rationale for using such a tool stems from the need to identify and select which operational barrier elements to prioritize in further analysis and follow-up activities as part of HF engineering or during operations.

Although several tools are available in freely accessible guidelines, attempts to apply these tools to operational barrier elements revealed that they were suboptimal for such use. They either proved to be too complex or targeted at specific systems and operations, or overly simplified and coarse in how they ranked criticality. While acknowledging that task criticality is a complex phenomenon, efforts were put into developing a tool which was as user friendly as possible, but without compromising on the ability to accurately distinguish more and less critical operational barrier elements. Definitions of what characterizes safety critical tasks were therefore developed and further transformed into a "roadmap" for determining criticality for operational barrier elements.

The presented tool (i.e. roadmap) is qualitative and intended to be flexible in its use to allow for expert judgement. Numerical scoring is not required and documenting the justification behind the criticality ranking is considered irrelevant, perhaps with the exception for highly critical operational barrier elements which may require substantial resources as part of further work. Experience from practical applications of the tool (Ludvigsen et al., 2018) shows that it is successful at distinguishing between operational barrier elements with high, medium and low criticality in what is considered a sensible distribution. While the tool should also be applicable to other types of safety critical tasks than operational barrier elements, adjustments may have to be done for it to be suitable for tasks with risk of introducing latent failures into the system, so called Type A actions. However, an argument is made about how approaches such as safety critical task analysis (or similar processes) may not be the most useful for managing such risks. It is instead encouraged to integrate HF principles and methods into other types of safety studies and risk analysis, such as HAZOP, LOPA and FMECA. Such techniques are intended for complete reviews of systems and can identify issues related to safety critical tasks with minor additional efforts. This will allow for time consuming and to some degree overlapping HF analyses to be skipped, and resources can instead be spent on implementation of measures targeted at improving design, procedures, training and other factors influencing human performance. As such, international standards such as IEC 61882 and IEC 60812 (among others) should be examined to check whether they could provide guidance on how to address HF issues as an integrated part of the safety study.

REFERENCES

Bridges, W. (2011). LOPA and Human Reliability – Human Errors and Human IPLs (Updated). *Prepared for Presentation at American Institute of Chemical Engineers 2011 Spring Meeting 7th Global Congress on Process Safety Chicago*, Illinois March 13–16, 2011.

Bye, A., Laumann, K., Taylor, C., Rasmussen, M., Øie, S., van de Merwe, K., Øien, K., Boring, R., Paltrinieri, N., Wærø, I., Massaiu, S., & Gould, K. (2017). *The Petro-HRA guideline.* Series: IFE/HR;E-2017/001. Retrieved via: www.ife.no

Center for Chemical Process Safety (2015). *Guidelines for initiating events and independent protection layers in layer of protection analysis.* Wiley: New York, NY.

Energy Institute (2011). *Guidance on human factors safety critical task analysis.* Energy Institute: London.

Gertman, D., Blackman, H., Marble, J., Byers, J., & Smith, C. (2005). *The SPAR-H human reliability analysis method* (NUREG/CR-6883). Retrieved via: www.nrc.gov

Health and Safety Executive (1999). *Human factors assessment of safety critical tasks.* Offshore Technology Report – OTO 1999 092. Retrieved via: www.hse.gov.uk

Health and Safety Executive (2016). *Offshore safety directive regulator: assessment principles for offshore cases (APOSC).* Retrieved via: www.hse.gov.uk

Ludvigsen, J.T., van de Merwe, K., Klemsdal le-Borgne, E., & Teigen, T. (2018). Applying an operational safety barrier framework in a major oil and gas field development project. *Safety and reliability – safe societies in a changing world.* Haugen, S., Barros, A., van Guliik, C., Kongsvik, T., & Vinnem, J.E. (Eds). Taylor & Francis Group, London.

Myers, P.M. (2013). Layer of protection analysis – quantifying human performance in initiating events and independent protection layers. *Journal of Loss Prevention in the Process Industries.* 26 (2013) 534–546.

NOPSEMA (2020). *Information paper: human factors – procedures and instructions.* Document No: N-06300-IP1041 A392397. Retrieved via: www.nopsema.gov.au

Øie, S., Wahlstrøm, A., Fløtaker, H.P., & Rørkjær, S. (2014). *Barrier management in operation for the rig industry: Good practices.* DNV GL Report No. 2013-1622. Rev. 1.

Petroleum Safety Authorities Norway (2013). *Prinsipper for barrierestyring i petroleumsvirksomheten.* Retrieved via: www.ptil.no

Petroleum Safety Authority Norway (2017). *Principles for barrier management in the petroleum industry: barrier Memorandum.* Retrieved via: www.ptil.no

Reason, J. (1997). *Managing the risks of organizational accidents.* Taylor & Francis, London and New York.

U.S. Chemical Safety and Hazard Investigation Board (2016). *Investigation report executive summary. Drilling rig explosion and fire at the Macondo well.* Report No. 2010-10-I-OS. Retrieved via: www.csb.gov

4 Making Sense of Sensemaking in High-Risk Organizations

B. E. Danielsen
NTNU

CONTENTS

INTRODUCTION

Safety in high-risk organizations is created by the everyday behavior of all employees in the organization – at all levels – as they go about getting their job done (Gregory & Shanahan, 2017). Sensemaking – the fundamental human quest for meaning – is the basis for human behavior, in formal organizations as well as life in general (Weick, Sutcliffe, & Obstfeld, 2005). Sensemaking has been an influential perspective in

53

organization studies and is strongly associated with the work of Karl Weick and his change of focus from decision-making and organizational outcomes to how individuals and organizations make sense of or give meaning to events and experiences (Weick, 1995). Sensemaking research has intensified over the last decades; however, it has been scattered over several different domains with differing approaches. In relation to the extensive literature on sensemaking, this chapter is a brief review of the sensemaking literature to establish an understanding of the concept, especially how it is understood in the context of sharp-end operators in high-risk organizations. Mainly, work on the bridge of a maritime vessel is used as an example to illustrate sensemaking in a high-risk environment.

This chapter is structured as follows: The next section presents the overall concept of sensemaking, a definition of sensemaking and how it relates to decision-making and situation awareness (SA). Thereafter, the third section describes how sensemaking is understood as a cognitive process, followed by the fourth section describing some of the debated core aspects of the sensemaking concept. A review of the main factors influencing sensemaking is presented in fifth section. The sixth section discusses sensemaking in safety-critical situations before the last section concludes with a summary and thoughts about future research opportunities.

THE CONCEPT OF SENSEMAKING

Sensemaking seems self-explanatory as it literally means "the making of sense"; however, as a cognitive concept, it reaches beyond merely being another word for "understanding" or "interpreting" (Weick, 1995). Interpretation implies there is something in the environment to be discovered; however, sensemaking is not a passive diagnosis, it refers to the processes where people actively play a role in constructing the very situations they try to make sense of (Weick, 1995). "Sensemaking is about sizing up a situation while you simultaneously act and partially determine the nature of what you discover" (Weick & Sutcliffe, 2015, p. 32).

Sensemaking both precedes and follows decision-making (Maitlis, 2005). Sensemaking is about "the interplay of action and interpretation rather than the influence of evaluation on choice" (Weick et al., 2005, p. 409). Snook (2000) effectively describes how sensemaking differs from decision-making in his analysis of a friendly fire incident over Iraq in 1994 where two US Air Force F-15 fighters accidentally shot down two helicopters killing all 26 peacekeepers on board. Snook points out that the F-15 pilots did not "decide" to pull the trigger. They were trying to make sense of the situation they were in. Ambiguous stimuli and strong expectations made the pilots believe they saw an enemy helicopter, "seeing through the mind's eye" as Snook puts it. Blaming the pilots for making the wrong decision would mean overlooking the "potent situation factors that influence action. Framing the individual-level puzzle as a question of meaning rather than deciding shifts the emphasis away from individual decision makers toward a point somewhere 'out there' where context and individual action overlap" (Snook, 2000, p. 207). Snook emphasizes that the individual sensemaking occurs as an interplay with the environment or embedded in context. To view human actions as a struggle to make sense rather than decision-making makes way for a more complete account for all relevant factors contributing to an accident, not merely individual judgment or "human error". It promotes a view of humans as "good people struggling to make sense" rather than "bad ones making poor decisions" (Snook, 2000, p. 207).

Sensemaking also differs from the widely used concept of SA (Endsley, 1995). SA is an individually achieved state of knowledge, based on the perception of elements in the environment and the comprehension of their meaning, which is used to make predictions about the future (Endsley, 1995). In contrast, the study of sensemaking is about the process of achieving these kinds of outcomes. Where SA seems to describe a more passive perception of data, sensemaking is about how people actively construct what counts as data in the first place (Klein, Phillips, Rall, & Peluso, 2007). The SA construct has been important within human factors research; it has been widely studied across many different domains. Still, it has been regarded as problematic that missing cues or displayed information is commonly described as "loss of situation awareness". This way of representing SA as a construct in the mind has made it possible to blame sharp-end operators for mishaps because they made the mistake of losing their SA (Dekker, 2015).

Sensemaking Definition

There is no unified definition of sensemaking. In the research literature it is often used without an associated definition, and when explicitly defined it is given a variety of meanings (Maitlis & Christianson, 2014). In their comprehensive literature review, Maitlis and Christianson (2014, p. 67) developed a definition of sensemaking rooted in recurrent themes across sensemaking definitions:

> a process, prompted by violated expectations, that involves attending to and bracketing cues in the environment, creating intersubjective meaning through cycles of interpretation and action, and thereby enacting a more ordered environment from which further cues can be drawn.

The various aspects on sensemaking from this definition will be discussed below.

THE SENSEMAKING PROCESS

Sensemaking understood as a cognitive process has been described as consisting of three interrelated processes: *creation, interpretation and enactment* (Sandberg & Tsoukas, 2015). The creation process involves noticing and extracting *cues* from our lived experience, creating an initial sense of the situation. In the interpretation process, the initial sense is then interpreted to a more complete and narratively organized sense of the situation. Following the interpretation process, the enactment process involves acting on the sense made. The actions create a slightly different or new environment to continue to make sense of (Maitlis & Christianson, 2014; Sandberg & Tsoukas, 2015). This meaning-action process is an ongoing cycle and sensemaking never starts or stops as people are in an "almost infinite stream of events and inputs" (Weick et al., 2005, p. 411).

TRIGGERS FOR SENSEMAKING

What makes something from this stream of events noticed and carved out as a cue for sensemaking? The sensemaking literature finds that issues, events or situations become triggers for sensemaking when they are ambiguous, interrupt people's ongoing activity, make them realize the inadequacy of their current understanding and create uncertainty about how to act (Klein et al., 2007; Maitlis & Christianson, 2014). The events triggering organizational sensemaking range from minor to major events and they may be planned or unplanned (Sandberg & Tsoukas, 2015). A triggering event does not necessarily emerge unexpectedly, it may be constructed by the actors themselves, e.g. by noticing or failing to notice cues (Ibid.). Cues are "seeds from which people develop a larger sense of what may be occurring" (Weick, 1995, p. 50).

Sometimes vague cues are not noticed, other times cues are significant enough to be noticed but still do not trigger sensemaking. Instead, they are accommodated, explained away or normalized (Maitlis & Christianson, 2014). How it makes sense to explain away cues is thoroughly described in Lützhöft and Dekker's (2002) analysis of the grounding of the *Royal Majesty* east of Nantucket in 1995. The ship lost her satellite signals just after departure from Bermuda and the GPS forwarded estimated positions to the autopilot. This went unnoticed by the crew as the automated bridge functions supported their mental model of a safe trip following the planned track to Boston. The crew explained away or did not attend to cues emerging along the way that would indicate a different story, like lookout reports and warning broadcasted on VHF. On June 10, 1995 she ran aground 17 nautical miles from her planned track towards Boston.

The study of sensemaking has mainly been confined to study episodes where an ongoing activity has been interrupted and need to be restored (Sandberg & Tsoukas, 2015). According to Sandberg and Tsoukas (2015), this is problematic as specific episodes certainly are an aspect of organizational life, but most of the time organizational life consists of routine work where people do things without deliberately thinking about how they do them. This does not mean that routine work is senseless, rather that people are involved in mundane or *immanent* sensemaking (Sandberg & Tsoukas, 2015). This is sensemaking as people are immersed in practice without being consciously aware of it, and they spontaneously respond to the situation as it develops. Sandberg and Tsoukas (2015) argue that the study of immanent sensemaking is as a way forward to extend the sensemaking concept. This line of enquiry would correspond to the safety II perspective on safety where the focus is on the everyday performance where things go right under varying conditions (Hollnagel, Woods, & Leveson, 2006).

SENSEMAKING AND ENACTMENT

Enactment is not only a stage in the sensemaking process but at the very core of the sensemaking concept (Sandberg & Tsoukas, 2015; Weick, 1988, 1995). Action is an integral part of sensemaking as it is a part of gathering more information about the situation at hand. Action can test the initial sense made as well as shape the environment for sensemaking. "What the world is without our enacting is never known since we fiddle with that world to understand it" (Weick pers. comm. cited in Klein et al., 2007, p. 122).

Sensemaking in a developing crisis can both be helpful and harmful as action can alter the environment in unexpected ways. Crises and unexpected events are situations difficult to comprehend, and the situation may require people to take action with incomplete information (Maitlis & Christianson, 2014). Enactment is central in Weick's analysis of a Union Carbide gas leak that occurred in Bhopal in 1984. Early action may determine the trajectory of the crisis, "Had they not acted or had they acted differently, they would face a different set of problems, opportunities and constraints" (Weick, 1988, p. 309). Enactment is especially difficult in complex systems where changes in one part have less predictable effect on other parts, there may be a delay in the effects of action and small actions can result in big and surprising effects (Maitlis & Christianson, 2014).

THE OUTCOME OF THE SENSEMAKING PROCESS

The outcome of the sensemaking process is "a more ordered environment from which further cues can be drawn" (Maitlis & Christianson, 2014, p. 67). The specific sense made is seen as a springboard for the actions people take to attempt to restore an interrupted activity. However, the sense made does not need to be an accurate account of the situation at hand. According to Weick "perceptual accuracy is grounded in models of rational decision-making" (Weick, 2005, p. 415). Sensemaking is not about discovering the truth and achieving a correct understanding, it is about "continuing redrafting of an emergent story" and it is "driven by plausibility rather than accuracy" (Weick, 1995, p. 55). As such, the endpoint for sensemaking is not a full comprehension of the situation or system at hand, it is a dynamic process occurring in a dynamic, changing environment (Klein et al., 2007). The sensemaker only needs a plausible explanation or narrative sufficient enough to continue their activity (Weick, 1995): "To deal with ambiguity, interdependent people search for meaning, settle for plausibility, and move on" (Weick et al., 2005, p. 419).

ASPECTS OF SENSEMAKING

There are different fundamental assumptions about the sensemaking concept found in the literature. The temporal orientation has been a subject for debate as well as whether sensemaking is primarily an individual cognitive process or a social construction of intersubjective meaning where language is the locus of sensemaking (Maitlis & Christianson, 2014).

SENSEMAKING AND TEMPORAL ORIENTATION

Weick (1995) listed *retrospective* as one of the seven distinct properties of sensemaking. Sensemaking rationalizes what people have done as they look back on action that has already taken place (Maitlis & Christianson, 2014; Weick, 1995). According to Weick people can know what they are doing only after they have done it. "The creation of meaning is an attentional process, but it is attention to that which has already occurred" (Weick, 1995, p. 26). Changes or cues in the environment are noticed when looking back over previous experience and seeing a pattern (Weick, 2005). As part of retrospective sensemaking, forward looking sensemaking has been explained as "future perfect" thinking where a future event is imagined and made sense of as if it had already occurred. However, in recent years, researchers have argued that sensemaking can also be prospective or future oriented (Gephart, Topal, & Zhang, 2011; Klein, Wiggins, & Dominguez, 2010; Rosness, Evjemo, Haavik, & Wærø, 2016) or be seen as drawing on all three dimensions (past, present and future) of sensemaking (Maitlis & Christianson, 2014).

Sensemaking and Language

Over time Weick has developed the notion of sensemaking in a way that gradually removed it from its cognitivist origins into a social constructivist perspective (Sandberg & Tsoukas, 2015). In this perspective, sensemaking is understood as being more fundamentally concerned with language (Maitlis & Christianson, 2014; Weick et al., 2005). The focus on language or linguistic factors has increased over the last two decades (Sandberg & Tsoukas, 2015). Language is a central part of organizational life as most social contact is mediated through talk and conversations. "Situations, organizations, and environments are talked into existence" (Weick et al., 2005, p. 409) and turning the flow of organizational circumstances into words and categories is central in sensemaking (Weick et al., 2005). Some scholars especially highlight narratives as the primary site from where experiences are made meaningful. Narratives are used to define individual and collective identities, and there may be several different narratives existing in an organization which contributes to people interpreting differently experiences they have in common (Brown, Stacey, & Nandhakumar, 2008; Maitlis & Christianson, 2014). Although the focus on language is connected to understanding sensemaking as the construction of intersubjective meaning rather than primarily as an individual cognitive process (Maitlis & Christianson, 2014), sensemaking can be understood as both an individual and a social process. According to Weick (1995) "sensemaking is grounded in both individual and social activity", and it might not even be possible to separate the two.

FACTORS INFLUENCING SENSEMAKING

As "people can make sense of anything" (Weick, 1995, p. 49), there is an infinite number of factors that can influence sensemaking. In this section, some of the central factors from the sensemaking literature is reviewed; emotions, embodied sensations, social context, identity, technology and meaningfulness through work (Maitlis & Christianson, 2014; Maitlis & Sonenshein, 2010; Sandberg & Tsoukas, 2015; Weick, 1995).

Sensemaking and Expectations

Sensemaking has often been described as a response to a surprise – a failure of expectations (Klein et al., 2007). Expectations can be both enabling or constraining for sensemaking (Weick, 1995). The discrepancy between expectations and reality must be of a certain magnitude or importance to cause people to wonder what is going on and trigger sensemaking (Maitlis & Christianson, 2014).

It can either be an unexpected event or the non-occurrence of an expected event. The experience of how significant a discrepancy feels is highly subjective; it can depend on the "impact on individual, social, or organizational identity ... and personal or strategic goals" (Maitlis & Christianson, 2014, p 70). It can also vary from moment to moment depending on emotions and identity construction (Weick, 1995).

In Weick's (1993) analysis of the Mann Gulch fire where 13 firefighters died, the firefighters expected a "10:00 fire", which meant it would be relatively easy to manage and be under control by 10:00 the next morning. This image stuck with them and prevented them from making sense of new cues as they emerged. Expectations were also an important part of the individual-level analysis of a friendly fire incident by Snook (2000). The F-15 fighter pilots had not been informed about the friendly helicopters when they entered what was designated as a "combat zone". Due to range, angle and speed of the fighters, the visual stimulus was ambiguous. As they flew close to the friendly helicopters a second time to confirm the sighting, they saw what they expected to see – the enemy.

SENSEMAKING AND EMOTIONS

Extensive research has found that the interplay between emotions and cognition influence who we are, what we do and the decisions we make (Norman, 2019). Emotions were initially ignored in sensemaking studies but have gradually been expanding in the recent years (Sandberg & Tsoukas, 2015). On both individual and collective levels, emotions have increasingly been understood to influence the sensemaking process, "whether sensemaking occurs, the form it takes, when it concludes, and what it accomplishes" (Maitlis & Christianson, 2014, p. 100).

Emotions in crisis are often strong and negative like anxiety fear, panic and desperation. The arousal these emotions trigger in the autonomous nervous system can consume cognitive information processing capacity, which in turn reduces the number of cues that can be noticed and become triggers for sensemaking (Maitlis & Sonenshein, 2010). As seen in the Mann Gulch incident when people are put under life-threatening pressure, they return to well-learned, habituated ways of responding, like flight (Weick, 1993). Positive felt emotions may "broaden individuals' scope of attention and their thought–action repertoires" (Maitlis & Sonenshein, 2010, p. 568). This should lead to a sensemaking process that can contribute to positive outcomes, averting crisis and accidents. However, overly positive emotions may cause people to be overly optimistic and overlook important cues and misinterpret the situation (Maitlis & Sonenshein, 2010). Hence, moderately intense emotions, strong enough to be noticed but not to distract and consume cognitive resources, seem to support sensemaking (Maitlis & Sonenshein, 2010).

EMBODIED SENSEMAKING

As sensemaking has been conceptualized as a deliberate process confined to specific episodes, research on sensemaking has mainly concerned the cognitive or linguistic sphere. However, over the recent years, focus on embodied sensemaking has emerged (Maitlis & Christianson, 2014). This research is connected to cognitive science and the related embodied cognition, where cognition, body, and context are viewed as three interrelated concepts that are in constant interaction with each other (Fahim & Rezanejad, 2014). Cunliffe and Coupland (2012) argued that "embodiment is an integral part of sensemaking" (Cunliffe & Coupland, 2012, p. 64) and that we make sense of our lives and ourselves through embodied interpretations of our ongoing everyday interactions and experiences. They theorized the process as "embodied narrative sensemaking". Through an analysis of rugby players, they demonstrated how the players made sense of their surroundings and experiences in sensory as well as cognitive ways; "sensemaking is not necessarily an information-processing activity but draws on an intuitive and informed feeling in his body – he senses the lines of force, the distance, his adversaries' positions on the field, and his critics off the field" (Cunliffe & Coupland, 2012, p. 77). They argue that we cannot separate our bodies from the context, in addition to the cognitive sphere "organizing also operates on a sensory level through sensory knowing and bodily sensations" (Cunliffe & Coupland, 2012, p. 83).

Embodied sensemaking is interesting in a maritime context where navigating a ship involves working in a highly dynamic environment. The form of tacit knowledge needed to maneuver a ship has been referred to as ship sense (Prison, Dahlman, & Lundh, 2013). Ship sense is presumed to play an important role in the dynamic interaction between the ship and the navigator. The navigator must handle the ship's distinctive maneuverability and the navigation instruments available, as well as account for the dynamic factors such as wind, waves, current and visibility that affect each other and the ship. When sailing in open sea with strong winds and high sea-state, the autopilot may be deliberately disengaged in order to steer the ship manually, and the more implicit knowledge to "get a feel for" the ship's movement becomes important. Both visual and other senses are engaged to feel the heaving motions of the vessel in the sea. Ship sense is needed to know when to take action, like slowing down or slightly altering course in relation to the direction of the oncoming waves (Prison et al., 2013), thus it is vital for the safety at sea.

Sensemaking Is Social

A lot of peoples' activities in organizations are concerned with collective efforts to make sense. Weick (1995) describes sensemaking as a social process where people actively shape each other's meanings. Sensemaking is never solitary as peoples' internal constructions or thoughts are created through interaction with others. In organizations "decisions are made either in the presence of others or with the knowledge they will have to be implemented, or understood, or approved by others" (Weick, 1995, p. 39). Sensemaking can thus be seen as unfolding between individuals as intersubjective meaning is constructed through a joint process of building understanding together (Maitlis & Christianson, 2014).

However, shared meaning is difficult to attain. People can share experiences although the sense made of it may differ significantly. For organizations, it is not even necessary that people share meanings to be able to coordinate action. It is sufficient to have minimal shared understanding or equivalent meanings (Brown et al., 2008;

Weick, 1995). Brown et al. (2008) were interested in why people interpret shared experiences differently. They explored the shared and discrepant sensemaking of members of a work team and argued that "although sensemaking is inherently social, it is fundamentally tied to processes of individual identity generation and maintenance" (Brown et al., 2008, p. 1037).

The social aspect of sensemaking is also understood by the sensemaking-related construct *sensegiving*, defined as "attempting to influence the sensemaking and meaning construction of others toward a preferred redefinition of organizational reality" (Gioia & Chittipeddi, 1991, p. 442). Organizational members may try to shape the sensemaking of others. Studies have shown that organizational leaders attempting to strategically shape the sensemaking of other organizational members do not necessarily succeed. Organizational members are not passive recipients of meaning; they engage in their own sensemaking and may actively resist the effort from leaders or alter the meanings conveyed to them (Maitlis & Christianson, 2014).

SENSEMAKING AND IDENTITY

Weick (1995) described sensemaking as being "grounded in identity construction" and that sensemaking begins with a self-conscious sensemaker. He argued that identities are constructed out of the process of interaction. "People learn about their identities by projecting them into an environment and observing the consequences" (Weick, 1995, p 23). Identity is thus not constant, as people experience a changing sense of self as they shift among interactions and try to decide which self is appropriate in the current situation (Weick, 1995). When the situation is ambiguous or confusing, sensemaking often occurs in ways that respond to people's identity needs (Weick, 2005). Sensemaking is part of maintaining a consistent, positive self-conception: "What the situation means is defined by who I become while dealing with it or what or who I represent" (Weick, 1995, p. 24). However, the direction of causality goes both ways: identity influences sensemaking but sensemaking also influences the definition of self.

The importance of identity for sensemaking becomes especially evident in organizational crises or change, when identity might be threatened (Maitlis & Sonenshein, 2010). A threatened identity may constrain action, as seen in Weick's (1993) analysis of the Mann Gulch fire. The foreman realized the severity of their situation and told the retreating crew to throw away their tools; however, without their tools they would turn "from a team of firefighters to a group of endangered individuals who were running from a fire without their tools" (Maitlis & Sonenshein, 2010, p. 563). Identity was also a contributing factor in the Westray mine disaster analyzed by Wicks (2001). Wicks found that institutionalization of a harmful mindset of invulnerability, e.g. they identified themselves as "real men", "going where few men would dare to go" (Wicks, 2001, p. 681), blinded them from seeing and preventing the risks in their work.

SENSEMAKING AND TECHNOLOGY

Sensemaking has been described as influenced by technology, particularly information and communication technologies; however, there are relatively few studies on this topic (Bisio, Bye, & Hurlen, 2019; Sandberg & Tsoukas, 2015). Organizational sensemaking is influenced by the medium of communication where people in organizations interact or the introduction of new technology triggers sensemaking about the technology itself or sensemaking related to professional identity (Sandberg & Tsoukas, 2015). As such, technology sensemaking has been treated as a subset of organizational sensemaking, focusing on sensemaking of the technological phenomenon in organizations rather than addressing how technological materiality influence sensemaking (Mesgari & Okoli, 2019).

There are sensemaking research strands mainly concerned with information seeking and the use of information technology. Sensemaking in the field of human–computer interaction (HCI) is concerned with tools for retrieving and visualizing information, how people make sense of complex sets of information and their ability to create and shape external representations of knowledge (Russell, Stefik, Pirolli, & Card, 1993). In Library and Information Science (LIS), the central sensemaking activities are information seeking, processing, creating, and using (Dervin, 1998). Today web-based tools have enabled people to seek and access large amounts of information, thus, the LIS and HCI communities have seen the need to start converging on projects "to help people make sense of the information resources now available" (Russell, Convertino, Kittur, Pirolli, & Watkins, 2018, p 3). Although sensemaking in organizations, like sensemaking on the bridge of a vessel, also includes information seeking and the use of information technology, these research strands do fully consider the context; the many different technological applications as well as the broader sociotechnical work environment.

How technology is influencing sharp-end operator's sensemaking can, for instance, be observed in the maritime sector. There has been a steady increase in digitalized products, applications and services introduced to this domain. The role and tasks of navigators have gone from navigating the vessel by means of manual control to increasingly having the role as managers of automated systems (Lützhöft, Grech, & Porathe, 2011). The navigators work has changed to increasingly become more and more dependent on *representations* of the outside world, making sense of an increasingly digitalized context (Danielsen & Lamvik, 2019).

Despite the increase in digitalized products, advanced automated systems and sensors on ships introduced to increase safety, there is still a high number of accidents at sea. Although shipping is becoming safer every year, in terms of the number of ships lost (Porathe, Hoem, Rødseth, Fjørtoft, & Johnsen, 2018), the European Maritime Safety Agency reported 3174 casualties and incidents in 2018 alone (EMSA, 2019).

Already in the mid-1980s, scholars described the challenges emerging in the work cooperation between people and technology (Bainbridge, 1983; Morgan Jr, Glickman, Woodard, Blaiwes, and Salas, 1986). Morgan et al. (1986) found that the interaction between individuals in a navy bridge team was partly determined by interaction with machines and machine procedures which left too little room for communication and cooperation between the team members. They argued that standardized performance of tasks can weaken the mechanisms that create the dynamics and flexibility presumed to strengthen a team's capacity to handle uncertainty and ambiguity – the very core situations that trigger sensemaking.

There are many examples where the interaction with electronic navigation equipment contributes to incidents and accidents (Chauvin, Lardjane, Morel, Clostermann, & Langard, 2013; Nilsen et al., 2017). The grounding of Royal Majesty earlier in this chapter is one example. Another example is the grounding of the Spain-registered bulk carrier *Muros* in 2016, as it was on passage between Teesport, UK, and Rochefort, France (MAIB, 2017). Although the route was planned and monitored using the vessel's Electronic Chart Display and Information System (ECDIS), the "system and procedural safeguards intended to prevent grounding were either overlooked, disabled or ignored" (MAIB, 2017, p. 20). For example, the track over Haisborough Sands was not planned or checked in an appropriate scale chart and the audible alarm and guard zone was disabled. The report states that "The ECDIS on board Muros had not been used as expected by the regulators or equipment manufacturers" (MAIB, 2017, p. 22). The latter sentence demonstrates a gap between how regulators and equipment manufacturers imagine work on board a maritime vessel is performed and how the seafarers actually go about solving their daily tasks. As the design of maritime technology often lack usability (Lützhöft & Vu, 2018), it hampers rather than help the navigator's sensemaking.

Despite today's extensive knowledge about how design of technology influence work performance and safety, the human–technology interaction problems persist (Strauch, 2017). Part of the challenge in the maritime sector is the many stakeholders and processes involved in the design of a ship's bridge, like regulations, shipowners, classification companies, designers and equipment manufacturers (Johnsen, Kilskar, & Danielsen, 2019; Jones, 2009; Lützhöft & Vu, 2018; Meck, Strohschneider, & Brüggemann, 2009; Merwe, 2016).

Sensemaking and Meaningfulness through Work

The increasing digitalization and automation of work is a general trend in our society. In the maritime sector, the development from a being a navigating navigator to a monitoring operator of automated systems may have unintended consequences. Introduction of new technology, automated systems and increasing proceduralization of work are seen by experienced seafarers as "marginalisation of professional competence, skills and judgements" (Kongsvik, Haavik, Bye, & Almklov, 2020). Work is a central human activity and meaningful work is a fundamental human need (Yeoman, 2014). Does it make sense to have a job where you are reduced to a set of eyes and ears, where a particular sensory input should trigger the use of a particular procedure? What makes work meaningful? Sensemaking is also the tool for which people experience their work as meaningful. Individual's perceptions of the significance of their work, experiencing a sense of purpose through their work efforts, are contributing to the experience of meaningful work (Rosso, Dekas, & Wrzesniewski, 2010). In turn, the meaning of work influences work motivation, behavior and performance (Rosso, Dekas, & Wrzesniewski, 2010) which are all crucial for safety in organizations (Gregory & Shanahan, 2017). The increasing automation of the workplace not only causes problems like human out-of-the-loop (Endsley & Kiris, 1995), automation surprise (Sarter, Woods, & Billings, 1997) and other issues concerning human-automation collaboration, it is a safety issue, as well as an ethical issue of designing meaningful work for people.

SENSEMAKING IN SAFETY-CRITICAL SITUATIONS

The term safety-critical situation denotes situations that, if they go wrong, have a large potential for causing harm to people, property or the environment. In the organization literature, the term "crisis" is also commonly used. Weick describes crises as "low probability/high consequence events that threaten the most fundamental goals of an organization" (Weick, 1988, p. 305). Events like this place strong demands on sensemaking as they are often "characterized by ambiguity of cause, effect, and means of resolution" (Maitlis & Sonenshein, 2010, p. 554). Several examples of sensemaking in these situations have been given throughout this chapter.

The studies of sensemaking in crises have been on both sensemaking as it unfolds during a crisis and how sense is made of crises after they happened (Maitlis & Christianson, 2014). The latter often draws on public inquiry reports and other documents that "have constructed an account of what happened, why it happened, and who was responsible" (Maitlis & Sonenshein, 2010, p. 554). Public inquiries can say something about the shared sensemaking process after a crisis and they may enable organizational learning (Brown, 2004, 2005; Maitlis & Sonenshein, 2010).

Research on sensemaking during crisis has included a range of sectors, from the space sector (Dunbar & Garud, 2009; Stein, 2004) and the air force (Snook, 2000) to mining (Wicks, 2001), climbing (Kayes, 2004) and entertainment events (Vendelo and Rerup, 2009). Weick analyzed the Bhopal accident (Weick, 1988, 2010), the Tenerife air disaster (Weick, 1990), the Mann Gulch fire (Weick, 1993) and the medical disasters of Bristol Royal Infirmary (Weick & Sutcliffe, 2003). Maitlis and Sonenshein (2010) found that the two central themes underlying sensemaking in crisis and change conditions are *shared meanings* and *emotions*. The criticality of shared meanings can be illustrated by a recent example of breakdown in team sensemaking.

BREAKDOWN IN TEAM SENSEMAKING

The social aspect of sensemaking was discussed earlier in this chapter. Klein et al. (2010) discussed the social aspect as team sensemaking, defined as "the process by which a team manages and coordinates its efforts to explain the current situation and to anticipate future situations, typically under uncertain or ambiguous conditions" (Klein et al., 2010, p. 304). Klein describes it as a macrocognitive function as it is the team rather than individuals that perform the sensemaking. A successful outcome of the team sensemaking process is a collective understanding of the situation which accommodates decision-making. Klein et al. (2010) points out that team sensemaking is not new or different type of sensemaking, it is about "the coordination of the team members as they seek data, synthesise the data and disseminate inferences" (Klein et al., 2010, p. 304). Sensemaking at the team level requires additional coordination and is more difficult to accomplish than individual sensemaking. According to Klein et al. (2010), breakdown in team sensemaking may more often contribute to accidents than sensemaking at the individual level as "Most failures can be traced to a breakdown in team sensemaking where critical cues were ignored and the teams failed to synthesise the existing information". The latter sentence may be a good description of what happened on the bridge of the frigate *HNoMS Helge Ingstad* before it collided with the tanker *Sola TS* in Hjeltefjorden on November 8, 2018 (AIBN, 2019). The comprehensive report from the accident investigation board takes into account a broad set of factors contributing to the accident, like organizational factors, leadership, teamwork, training and technology, on the frigate and the other involved actors *Sola TS* and the Fedje VTS (AIBN, 2019). But as a case of breakdown in team sensemaking, it is interesting to take a look at what happened on the bridge of *HNoMS Helge Ingstad* minutes before the collision.

On the bridge of *HNoMS Helge Ingstad*, the structure of team sensemaking was hierarchical (Klein et al., 2010). In such a structure, the data should flow from different sources to a common node, in this case the officer of the watch (OOW), who puts the pieces together and directs the search for new data. The OOWs role was being responsible for conveying a clear and authorative picture of the situation (AIBN, 2019). To use Kleins' vocabulary, the OOW was the *data synthesizer*, which is a difficult task as the relevant information resides in different places.

During the watch handover sometime between 03:36 and 03:53, the OOW about to be relieved and the oncoming OOW observed an object (the tanker *Sola TS*) at the Sture Terminal starboard of the frigate's course line. It was observed both visually and on the radar display, however shown as an Automatic Identification System (AIS) signal without speed vector. During discussion between the two OOWs, they formed a clear perception (selected a frame) that the "object" was stationary near the shore. According to the data/frame theory of sensemaking (Klein et al., 2007) when a frame is selected it is used to guide further information seeking. During the handover, the first opportunity to gather further data was missed as they did not use the AIS to obtain more information about the "object". The relieving OOW's mental model was from this point very stable and the subsequent data-seeking and actions were based on his selected frame of reference.

Since the "object" was understood as stationary, it was not tracked by any of the radars; hence the bridge system did not generate any alarms to indicate that the vessel was on collision course with *Sola TS*. The "object" was primarily observed visually, and when *Sola TS* first started maneuvering out from the quay, this was done so slowly that it was difficult to perceive any movement from the bridge on *HNoMS Helge Ingstad*. Further visual observations by the bridge crew did not change the impression of the "object". None of the bridge team members saw the navigation lights on *Sola TS*; they only observed the strong deck lights. When the OOW saw that a little distance had appeared between the shore and the "object" on the radar, this cue was explained by assuming that the distance between the shore and the "object" on the radar screen was due to the frigate having come closer to the point which the "object" lay alongside.

The information the OOW received from the rest of the bridge team gave no indication that the "object" posed any risk to the voyage. The team's capacity to monitor the traffic situation was reduced due to a temporary unmanned starboard lookout position. In addition, the OOW and the bridge team were focusing on training activities and on the three vessels approaching in the opposite direction on the port side of

HNoMS Helge Ingstad. These vessels had been observed visually and tracked in the bridge system. Of those present, only the helmsman had identified the lights ahead on starboard side as belonging to a moving vessel. However, he did not disseminate this observation; he assumed that the OOW and the rest of the bridge team were aware of it being a vessel, since it should be observable on both AIS and radar. A minute before the collision, the pilot on board *Sola TS* made a direct call on VHF requesting *HNoMS Helge Ingstad* to alter course to starboard. The OOWs understanding was that there was not enough room to pass on the shore side of the "object" and assumed that the call was from one of the three northbound vessels approaching on port side. He responded by saying that they could not turn to starboard. By the time the OOW understood that the "object" giving off light was moving and on direct course to collide it was too late, and at 04:01:15, *HNoMS Helge Ingstad* collided with the tanker *Sola TS* (AIBN, 2019).

In hindsight it is easy to see that the cues and the weak signals of danger were there: the AIS signal, the radar echo, information on VHF and the visual information. However, at the time the selected mental frame of the OOW and the rest of the bridge team were used to guide information seeking. According to the data/frame theory of sensemaking (Klein et al., 2007), a surprise or an inadequacy in an existing frame will lead individuals to either actively obtain more relevant data to improve the frame, replace the frame with a more relevant one, construct a new frame or preserve the frame by explaining away or distorting the data. In this case the frame was preserved, and it directed expectations and what counted as data. The one team member that had made sense of the visual information was not able to share this information with the team and as such the team was "less sensitive to the weak signals than the most sensitive of their individual members" (Klein et al., 2010).

ADAPTIVE SENSEMAKING

As we have seen, sensemaking can be both helpful and harmful in safety-critical situations. Maitlis and Sonenshein (2010) argue that adaptive sensemaking is necessary for sensemaking to be helpful especially in crisis. Adaptive sensemaking is enabled when emotions are moderately intense, not too negative but not too positive either. Emotions can provide valuable information to a sensemaker and should be intense enough to be noticed. However, the capacity for anxiety toleration is important for the ability to make sense of a situation (Stein, 2004).

According to Maitlis and Sonenshein (2010), another enabler of adaptive sensemaking is the two processes of *updating* and *doubting*. Updating has to do with gathering new information and revising interpretations while doubt is a reminder of constantly generating new understandings. A finite sense of a situation is never made as things are always changing (Maitlis & Sonenshein, 2010).

Adaptive sensemaking is related to improvisation. The ability to improvise has been connected to resilience (Weick & Sutcliffe, 2015). Skill in improvisation, as well as having the flexibility to use it, increases the potential actions available in people's repertoire, meaning they can act on a greater variety of situations and surprises as well as broaden the range of cues that can be noticed (Ibid). The connection between sensemaking and resilience has been made by several scholars, in the sense that sensemaking creates resilience but also that sources of resilience help to make sense of the situation (Kilskar, Danielsen, & Johnsen, 2020).

CONCLUSION

This chapter reviewed sensemaking literature and defined sensemaking as "a process, prompted by violated expectations, that involves attending to and bracketing cues in the environment, creating intersubjective meaning through cycles of interpretation and action, and thereby enacting a more ordered environment from which further cues can be drawn" (Maitlis & Christianson, 2014, p. 67). The cognitive sensemaking process of creation, interpretation and enactment is described, highlighting that sensemaking is about partly creating the environment or situation to make sense of (Weick, 1995). Sensemaking is most often described as being triggered by ambiguous or surprising events influenced by expectations, emotions, embodied sensations, technology, the social context for sensemaking as well as identity. Safety-critical situations or crisis are especially demanding for sensemaking. Sensemaking should be adaptive to be helpful in these situations (Maitlis & Sonenshein, 2010).

In the context of sociotechnical systems, the sharp end operators, like navigators, are confronted with dynamic evolving situations and complex stimuli. According to Klein et al. (2007), sensemaking contributes to our understanding of human behavior at a macrocognitive scale needed to understand and design complex cognitive systems. As such it should be a useful perspective for both accident analysis and for the future design of safety-critical systems.

There are several areas where future research could develop the sensemaking concept or perspective further (Maitlis & Christianson, 2014; Sandberg & Tsoukas, 2015). In line with Sandberg and Tsoukas (2015), this review found that there are relatively few studies investigating how technologies influence sensemaking, especially how the design of technology can hamper or support sensemaking in high-risk industries, like the maritime sector. The design of technology is part of a general trend with increasing digitalization and automation of work. An interesting direction for future sensemaking research would be to investigate how meaningful work can be designed in this context and how meaningfulness relates to resilience and safety.

REFERENCES

AIBN. (2019). *Report on the collision on 8 November 2018 between the frigate HNoMS Helge Ingstad and the oil tanker Sola TS outside the Sture terminal in the Hjeltefjord in Hordaland county.* Accident Investigation Board Norway

Bainbridge, L. (1983). Ironies of automation. *Automatica, 19*(6), 775–779.

Bisio, R., Bye, A., & Hurlen, L. (2019). Human machine interface for supporting sensemaking in critical situations, a literature review. *Paper presented at the European Safety and Reliability Conference, Hannover, Germany.*

Brown, A. D. (2004). Authoritative sensemaking in a public inquiry report. *Organization Studies, 25*(1), 95–112. doi:10.1177/0170840604038182

Brown, A. D. (2005). Making sense of the collapse of Barings Bank. *Human Relations, 58*(-12), 1579–1604. doi:10.1177/0018726705061433

Brown, A. D., Stacey, P., & Nandhakumar, J. (2008). Making sense of sensemaking narratives. *Human Relations, 61*(8), 1035–1062. doi:10.1177/0018726708094858

Chauvin, C., Lardjane, S., Morel, G., Clostermann, J.-P., & Langard, B. (2013). Human and organisational factors in maritime accidents: Analysis of collisions at sea using the HFACS. *Accident; Analysis and Prevention, 59*, 26–37.

Cunliffe, A., & Coupland, C. (2012). From hero to villain to hero: Making experience sensible through embodied narrative sensemaking. *Human Relations, 65*(1), 63–88.

Danielsen, B.-E., & Lamvik, G. (2019). Making sense of bridge design: how seamanship may challenge technology-as-designed. *Paper presented at the Proceedings of the 29th European Safety and Reliability Conference (ESREL). 22–26 September 2019 Hannover, Germany.*

Dekker, S. W. (2015). The danger of losing situation awareness. *Cognition, Technology & Work, 17*(2), 159–161.

Dervin, B. (1998). Sense-making theory and practice: an overview of user interests in knowledge seeking and use. *Journal of Knowledge Management, 2*(2), 36–46.

Dunbar, R. L. M., & Garud, R. (2009). Distributed knowledge and indeterminate meaning: The case of the Columbia shuttle flight. *Organization Studies, 30*(4), 397–421. doi:10.1177/0170840608101142.

EMSA. (2019). *European maritime safety agency annual overview of marine casualties and incidents 2019.* Retrieved from http://www.emsa.europa.eu/emsa-documents/latest/item/3734-annual-overview-of-marine-casualties-and-incidents-2019.html

Endsley, M. R. (1995). Toward a theory of situation awareness in dynamic systems. *Human Factors, 37*(1), 32–64.

Endsley, M. R., & Kiris, E. O. (1995). The out-of-the-loop performance problem and level of control in automation. *Human Factors, 37*(2), 381–394.

Fahim, M., & Rezanejad, A. (2014). An introduction to embodied cognition. *International Journal of Language and Linguistics, 2*(4), 283–289.

Gephart, R. P., Jr., Topal, C., & Zhang, Z. (2011). Future-oriented sensemaking: temporalities and institutional legitimation. In *Process, Sensemaking, and Organizing.* Oxford University Press.

Gioia, D. A., & Chittipeddi, K. (1991). Sensemaking and sensegiving in strategic change initiation. *Strategic Management Journal, 12*(6), 433–448. doi:10.1002/smj.4250120604.

Gregory, D., & Shanahan, P. (2017). *Being Human in Safety-Critical Organisations.* Norwich, UK: TSO.

Hollnagel, E., Woods, D. D., & Leveson, N. (2006). *Resilience Engineering: Concepts and Precepts.* Abingdon: Ashgate Publishing Ltd.

Johnsen, S. O., Kilskar, S. S., & Danielsen, B.-E. (2019). Improvements in rules and regulations to support sensemaking in safety-critical maritime operations. *Paper presented at the Proceedings of the 29th European Safety and Reliability Conference (ESREL). 22–26 September 2019 Hannover, Germany.*

Jones, B. S. (2009). Unpacking the blunt end–changing our view of organizational accidents. Paper presented at the Contemporary Ergonomics 2009: Proceedings of the International Conference on Contemporary Ergonomics 2009.

Kayes, D. C. (2004). The 1996 Mount Everest climbing disaster: The breakdown of learning in teams. *Human Relations, 57*(10), 1263–1284. doi:10.1177/0018726704048355

Kilskar, S. S., Danielsen, B.-E., & Johnsen, S. O. (2020). Sensemaking in critical situations and in relation to resilience—A review. *ASCE-ASME Journal of Risk and Uncertainty in Engineering Systems, Part B Mechanical Engineering, 6*(1).

Klein, G., Phillips, J. K., Rall, E. L., & Peluso, D. A. (2007). A data-frame theory of sensemaking. *Paper presented at the Expertise Out of Context: Proceedings of the Sixth International Conference on Naturalistic Decision Making.*

Klein, G., Wiggins, S., & Dominguez, C. O. (2010). Team sensemaking. *Theoretical Issues in Ergonomics Science, 11*(4), 304–320. doi:10.1080/14639221003729177

Kongsvik, T., Haavik, T., Bye, R., & Almklov, P. (2020). Re-boxing seamanship: From individual to systemic capabilities. *Safety Science, 130*, 104871.

Lützhöft, M., Grech, M. R., & Porathe, T. (2011). Information environment, fatigue, and culture in the maritime domain. *Reviews of Human Factors and Ergonomics, 7*(1), 280–322.

Lützhöft, M., & Vu, V. D. (2018). Design for safety. In H. A.Oltedal and M. Lützhöft (Eds.), *Managing Maritime Safety*. New York: Routledge.

Lützhöft, M. H., & Dekker, S. W. A. (2002). On your watch: Automation on the bridge. *Journal of Navigation, 55*(1), 83–96. doi:10.1017/S0373463301001588

MAIB. (2017). *Report on the investigation of the grounding of Muros Haisborough Sand North Sea 3 December 2016.* (22/2017). Marine Accident Investigation Branch UK

Maitlis, S. (2005). The social processes of organizational sensemaking. *Academy of Management Journal, 48*(1), 21–49. doi:10.5465/AMJ.2005.15993111

Maitlis, S., & Christianson, M. (2014). Sensemaking in organizations: Taking stock and moving forward. *The Academy of Management Annals, 8*(1), 57–125.

Maitlis, S., & Sonenshein, S. (2010). Sensemaking in crisis and change: Inspiration and insights from weick (1988). *Journal of Management Studies, 47*(3), 551–580. doi:10.1111/j.1467–6486.2010.00908.x

Meck, U., Strohschneider, S., & Brüggemann, U. (2009). Interaction design in ship building: An investigation into the integration of the user perspective into ship bridge design. *Journal of Maritime Research, 6*(1), 15–32.

Merwe, F. V. D. (2016). *Human-centred design of alert management systems on the bridge.* Retrieved from https://www.sdir.no/sjofart/ulykker-og-sikkerhet/sikkerhetsutredninger-og-rapporter/human-centred-design-of-alert-management-systems-on-the-bridge/

Mesgari, M., & Okoli, C. (2019). Critical review of organisation-technology sensemaking: Towards technology materiality, discovery, and action. European Journal of Information Systems, 28(2), 205–232.

Morgan, B. B. Jr, Glickman, A. S., Woodard, E. A., Blaiwes, A. S., & Salas, E. (1986). *Measurement of team behaviors in a navy environment.* Human Factor Division, Naval Training Systems Center, Department of The Navy Battelle Columbus Labs Research Triangle Park, NC.

Nilsen, M., Almklov, P.G., Haugen, S., & Bye, R.J. (2017). A discussion of risk influencing factors for maritime accidents based on investigation reports. *Paper presented at the ESREL, Glasgow, Scotland.*

Norman, E. (2019). *Affekt og kognisjon.* Oslo: Universitetsforlaget.

Porathe, T., Hoem, Å., Rødseth, Ø., Fjørtoft, K., & Johnsen, S. O. (2018). At least as safe as manned shipping? Autonomous shipping, safety and "human error". In Haugen (Ed.), *ESREL European Safety and Reliability Conference* (pp. 417–425). London: Taylor & Francis Group.

Prison, J., Dahlman, J., & Lundh, M. (2013). Ship sense-striving for harmony in ship manoeuvring. *WMU Journal of Maritime Affairs, 12*(1), 115–127.

Rosness, R., Evjemo, T. E., Haavik, T., & Wærø, I. (2016). Prospective sensemaking in the operating theatre. *Cognition, Technology and Work, 18*(1), 53–69. doi:10.1007/s10111-015-0346-y

Rosso, B. D., Dekas, K. H., & Wrzesniewski, A. (2010). On the meaning of work: A theoretical integration and review. *Research in Organizational Behavior, 30*, 91–127.

Russell, D. M., Convertino, G., Kittur, A., Pirolli, P., & Watkins, E. A. (2018). Sensemaking in a Senseless World: 2018 Workshop Abstract. *Paper presented at the Extended Abstracts of the 2018 CHI Conference on Human Factors in Computing Systems.*

Russell, D. M., Stefik, M. J., Pirolli, P., & Card, S. K. (1993). The cost structure of sensemaking. *Paper presented at the Proceedings of the INTERACT'93 and CHI'93 Conference on Human Factors in Computing Systems.*

Sandberg, J., & Tsoukas, H. (2015). Making sense of the sensemaking perspective: Its constituents, limitations, and opportunities for further development. *Journal of Organizational Behavior, 36*, S6-S32. doi:10.1002/job.1937

Sarter, N. B., Woods, D. D., & Billings, C. E. (1997). Automation surprises. *Handbook of Human Factors and Ergonomics, 2*, 1926–1943.

Snook, S. A. (2000). *Friendly fire: The accidental shootdown of U.S. Black Hawks over Northern Iraq.* Princeton, NJ: Princeton University Press.

Stein, M. (2004). The critical period of disasters: Insights from sense-making and psycho-analytic theory. *Human Relations, 57*(10), 1243–1261. doi:10.1177/0018726704048354

Strauch, B. (2017). Ironies of automation: Still unresolved after all these years. IEEE Transactions on Human-Machine Systems, 48(5), 419–433. doi:10.1109/THMS.2017.2732506

Vendelo, M. T., & Rerup, C. (2009). Weak cues and attentional triangulation: The Pearl Jam concert accident at Roskilde Festival. *Paper presented at the Academy of Management Annual Meeting, Chicago.*

Weick, K. E. (1988). Enacted sensemaking in crisis situations. *Journal of Management Studies, 25*(4), 305–317.

Weick, K. E. (1990). The vulnerable system: An analysis of the Tenerife air disaster. *Journal of Management, 16*(3), 571–593. doi:10.1177/014920639001600304

Weick, K. E. (1993). The collapse of sensemaking in organizations: The Mann Gulch disaster. *Administrative Science Quarterly, 38*(4), 628–652.

Weick, K. E. (1995). *Sensemaking in Organizations.* Thousand Oaks, California: SAGE.

Weick, K. E. (2005). Organizing and failures of imagination. *International Public Management Journal, 8*(3), 425–438. doi:10.1080/10967490500439883

Weick, K. E. (2010). Reflections on enacted sensemaking in the Bhopal disaster. *Journal of Management Studies, 47*(3), 537–550. doi:10.1111/j.1467–6486.2010.00900.x

Weick, K. E., & Sutcliffe, K. M. (2003). Hospitals as cultures of entrapment: A reanalysis of the Bristol Royal Infirmary. *California Management Review, 45*(2), 73. doi:10.2307/41166166

Weick, K. E., & Sutcliffe, K. M. (2015). *Managing the Unexpected: Sustained Performance in a Complex World.* New Jersey: John Wiley & Sons Inc.

Weick, K. E., Sutcliffe, K. M., & Obstfeld, D. (2005). Organizing and the process of sensemaking. *Organization Science, 16*(4), 409–421. doi:10.1287/orsc.1050.0133

Wicks, D. (2001). Institutionalized mindsets of invulnerability: Differentiated institutional fields and the antecedents of organizational crisis. *Organization Studies, 22*(4), 659–692. doi:10.1177/0170840601224005

Yeoman, R. (2014). Conceptualising meaningful work as a fundamental human need. *Journal of Business Ethics, 125*(2), 235–251.

5 Prospective Sensemaking in Complex Organizational Domains

A Case and Some Reflections

T. E. Evjemo
SINTEF

CONTENTS

INTRODUCTION

This chapter takes as a starting point the accident where the Norwegian frigate *KNM Helge Ingstad* collided with the oil tanker *Sola TS* in Hjeltefjorden. The aim of the chapter is to operationalize *prospective sensemaking* to lay the grounds for an improved understanding of events that led up to the collision. The tanker was sailing on an intersecting course, and the collision resulted in the frigate being evacuated prior to sinking. The collision occurred in an area where all ship traffic was monitored by a shipping control centre named Fedje VTS (Vessel Traffic Service). Afterwards, in the media and the public space, critical questions were soon raised about how such an accident could occur at all, including the publication of an audio log immediately afterwards where one hears the frigate, the tanker and the shipping control centre communicate. The Accident Investigation Board Norway (AIBN) published the first of two reports in late 2019 focusing on human aspects related to the accident. According to the AIBN report, misunderstanding of the situation on the frigate was the root cause while specific criticism is put forth towards both the tanker and the shipping control centre.

It is possible to look at the events in Hjelteforden as consisting of actors in a complex sociotechnical system – collaboration being imperative to ensure safety of the system. In such a system, people are essential when it comes to safeguarding safety. Such an approach involves analytical focus on what people do, when they do and how the actions affect other people. The argument is that to manage safety one must be able to understand *how* different actors' dependence on each other's actions plays out. Analytically, *interdependence* is therefore a key issue when analysing social interaction within complex organizational domains. It is also important to bear in mind that neither the frigate, the tanker nor the shipping control centre undertook anything that differed from the usual, which means that the accident occurred during routine and normal operations. An analysis of this type of accident is also about understanding how something that apparently goes well, paradoxically changes in a very short time span. Therefore, to learn from the *Helge Ingstad* accident and in particular make the findings relevant beyond this accident per se, the argument is that it is necessary to analyse the events prior to the collision itself in the context of an empirically grounded (safety) concept – an analysis this chapter facilitates through proposing how to operationalize prospective sensemaking.

The properties of prospective sensemaking were elaborated in the article by Rosness, Evjemo, Haavik and Wærø (2015), whereas the AIBN's investigation report from 2019 described the sequence and handling of events prior to the collision between *Helge Ingstad* and *Sola TS*. This chapter develops an operationalization of the properties of prospective sensemaking by exploring the unfolding events prior to the collision. The following two research questions are elaborated: 1. *How can prospective sensemaking be operationalized on the grounds of the KNM Helge Ingstad accident? 2. What can be learned from an applied operationalization of prospective sensemaking?*

MOTIVATION AND METHOD

The motivation behind operationalization of prospective sensemaking is twofold: firstly, to better understand how such an ordinary traffic pattern and normal situation could escalate into a collision with the dire consequences that followed. For this chapter, such an understanding involves a systemic approach to understanding collaboration, i.e., focusing on the actions the actors perform – more specifically coordinating and communicative practices and their enabling resources. Particularly, this chapter views technology use as rooted in local situations and action being in line with the work of Suchman (1987, 2007). Secondly, to explore whether our understanding of this type of accident can be improved by utilizing the concept of prospective sensemaking. The use of prospective sensemaking in this chapter is also related to traditional accident investigations, which often take as their starting point an exclusive individual, cognitive approach to human behaviour with an emphasis on what went wrong, which can be somewhat detrimental to understand also the social processes that are important when people cooperate and coordinate safety critical activities among themselves. In this context, the study by De Boer and Dekker (2017) is particularly intriguing. They have studied automation surprise within the airline cockpit, more specifically looking into two distinct perspectives with different approaches to pilots' cognitive processes regarding human–automation interaction. They argue that to improve safety, a sensemaking approach focusing on systemic factors including the operative context will make more sense compared to a traditional individualistic- and failure-centred focus on human performance.

However, this chapter goes beyond the traditional notion of sensemaking advocated by Weick (1995), Weick, Sutcliffe and Obstfeld (2005) and De Boer and Dekker (2017), drawing on the idea of *prospective sensemaking* which foremost implies sensemaking processes understood analytically as primarily forward oriented (Rosness et al., 2015). Furthermore, prospective sensemaking in this chapter will be discussed analytically in the context of how coordination per se becomes relevant among the actors in Hjeltejorden, i.e., prospective sensemaking does entail social aspects (Weick, 1995), but also needs a follow up via a distinct social-analytical focus.

The next section presents the events that unfolded prior to the collision between *Helge Ingstad* and *Sola TS* based on the AIBN (2019) report. Thereafter, an account is provided on traditional sensemaking, related concepts and the properties of prospective sensemaking followed by an account on how one can operationalize prospective sensemaking based on the AIBN (2019) report. The chapter concludes with reflections on what can be learned by an operationalization of prospective sensemaking within complex organizational domains.

THE FRIGATE, THE TANKER AND THE SHIPPING CONTROL CENTRE (VTS) – COURSE OF EVENTS PRIOR TO THE COLLISION

The following section presents a simplified, descriptive review of the course of events prior to the collision between *KNM Helge Ingstad* and *Sola TS* outside Stureterminalen in Hjelteforden in 2018. The review is based on the events as described in the AIBN (2019) report. The collision between the frigate *KNM Helge Ingstad* and the tanker *Sola TS* occurred overnight on Thursday 8 November 2018 in Hjeltefjord in the waters off the Sture terminal. The frigate crew consisted a total of 137 people, of which the bridge crew made up 7 people. Of these seven, two were undergoing training. *Sola TS* had a crew of 24 people where 4 people were on the bridge, including a pilot. The course of events in the AIBN report is mapped out by obtaining technical and electronic information from both vessels, from Fedje VTS including radio and radar recordings, as well as AIS log from the Norwegian Coastal Administration. Interviews have been conducted with key people involved, central documents were obtained, technical investigations aboard the frigate, and a similar voyage with a ship of the same frigate class, and with *Sola TS*.

The accident occurred at night when *KNM Helge Ingstad* sailed south in Hjeltefjord at a speed of just under 20 knots. The frigate did not send out an AIS signal as the automatic identification system was set to passive mode; however, the crew had earlier notified Fedje VTS about positioning and intentions via mobile phone. At the same time, the frigate listened to VHF channel 80, which is Fedje VTS' radio channel frequency. When the traffic controller at Fedje VTS became aware of the frigate, the ship was logged but not plotted into Fedje VTS' monitoring system. When the frigate was 17 nautical miles from the Sture terminal, *Sola TS* was about to leave the Sture terminal. On departure, the tanker had the navigational lanterns on but also part of the deck lighting.

According to the accident report, *Sola TS's* pilot reported the departure via VHF radio to Fedje VTS, after which the traffic controller acknowledged the call and zoomed in on the area around *Sola TS* on his screen. *KNM Helge Ingstad* was now about 6 nautical miles north of *Sola TS* and at the same time off the screen of the traffic controller at Fedje VTS. The screen continued to be zoomed in at the Sture terminal even after *Sola TS* departed. At the same time, a change of Officer On Watch (OOW) began on *KNM Helge Ingstad's* bridge – while the frigate was controlled by an officer under OOW training. After *Sola TS* left the Sture terminal, the tanker maintained a speed of about 7 knots on an easterly course in Hjeltefjord before initiating a turn towards the north, having both the navigational lanterns and the deck lights turned on. The accident report points out that none of the messages sent on VHF channel 80 from *Sola TS* to Fedje VTS were recorded by *KNM Helge Ingstad*. *Sola TS* was not tracked electronically by *KNM Helge Ingstad*, and at the same time the frigate radar was not used to monitor the surrounding waters. However, three other incoming vessels were tracked on the frigate bridge. As the frigate and the tanker approached each other, *KNM Helge Ingstad's* OOW changed course to port several times to increase the distance to the now-observed "object". At the same time, neither *KNM Helge Ingstad* nor any of the other vessels in Hjeltefjord were plotted on *Sola TS's* radar.

About 4 minutes before the collision, *Sola TS's* pilot responded to the object on the opposite course. The distance between the ships was now down to 4 nautical miles. *KNM Helge Ingstad* sailed without AIS signals, which meant that the identity of the frigate was not displayed on *Sola TS's* screens. The pilot contacted Fedje VTS for information on the oncoming vessel, but since the traffic controller had not recently monitored the movement of the frigate, it was not possible to confirm the identity of the frigate immediately. *Sola TS* then attempted to make contact using the Aldis lamp, which was unsuccessful. Eventually, the pilot at *Sola TS* received feedback from Fedje VTS about which ship was coming towards them – the pilot then tried to contact *KNM Helge Ingstad* but it took over 2 minutes before he succeeded. The pilot asked the OOW to swing starboard, something the frigate did not comply with. Consequently, when *KNM Helge Ingstad* did not change course, the machine was first stopped on board the *Sola TS*, then full aft. At this point, however, a collision was inevitable.

THE TRADITIONAL NOTION OF (RETROSPECTIVE) SENSEMAKING

The traditional approach to organizational sensemaking has involved foremost retrospective processes, i.e., provide meaning to what has already happened. One distinguishes between a cognitive approach to sensemaking, i.e., how individuals assign meaning to own experiences, and an organizational perspective which involves social dimensions or interactional processes between or among people. According to

Weick (1995), sensemaking is triggered by uncertainty or ambiguity, meaning situations involving both incomprehensibility as well as the possibility of several possible interpretations. The following quote is illustrative, "... explicit efforts at sensemaking tend to occur when the current state of the world is perceived to be different from the expected state of the world, or when there is no obvious way to engage the world" (Weick et al., 2005:409).

Moreover, we as actors cannot relate to an objective reality out there that can verify our efforts to make sense of something. Furthermore, *retrospective* means that our (attentional) creation of meaning, i.e., our sensemaking processes are targeted primarily on the past for us to construct meaning to what is happening now, or as Weick et al. (2005) describe, "... sensemaking involves the ongoing retrospective development of plausible images that rationalize what people are doing. Viewed as a significant process of organizing, sensemaking unfolds as a sequence in which people concerned with identity in the social context of other actors engage ongoing circumstances from which they extract cues and make plausible sense retrospectively, while enacting more or less order into those ongoing circumstances" (Weick et al., 2005:409). Sensemaking is thus directed backwards from any distinct situation, i.e., whatever is going on now will be impacted by what one experiences when looking backwards. It might be that trying to correctly identify a certain object in front is hampered due to having successfully identified (and got verification after some time) a similar object several times before in the same position and at the same moment in time. Arguably, our retrospective sensemaking can be biased.

At the same time, Weick's approach to sensemaking implies that sensemaking processes are put in motion only when people experience events or situations where meaning becomes precarious, or worse, lost. Paradoxically, this means that when everything goes smoothly or as planned, for example, when ship navigation proceeds as expected, Weick (1995) argues that sensemaking is not prominent per se, hence also challenging to identify. Arguably, the crew on *KNM Helge Ingstad* can be said to experience the sailing prior to the collision as "normal", in the sense that none on the bridge uttered any immediate major concerns regarding the course of events. Consequently, applying a traditional sensemaking perspective without anything being *unnormal* on the frigate bridge or within Hjeltefjorden per se would be somewhat challenging from Weick's position.

CONCEPTS RELATED TO SENSEMAKING

The concept of situation awareness (SA) is well known within safety science as well as various industries. Endsley (1995) defines SA as "... the perception of the elements in the environment within a volume of time and space, the comprehension of their meaning and the projection of their status in the near future" (Endsley, 1995:36). SA is a skill that has three cognitive phases: the first involving the ability to perceive elements in a situation. On a ship's bridge it can involve perceiving an increase in speed. Phase two is about comprehension of a given situation, i.e., perceived elements become meaningful in the current situation. A speed increase might be the result of new incoming information. The third phase implies projecting the future state based on previous experience. Moreover, the concept of SA is often talked about in relation to the term "loss of SA", which implies that SA is seen as something that people have, i.e., it is not a process per se (Dekker, 2013). However, within the genre computer-supported cooperative work (CSCW), novel technology is discussed on the grounds whether it can facilitate collaboration between actors. A classic approach to awareness within CSCW is Dourish and Bellotti's (1992), arguing that awareness implies an understanding of others' activities, which then acts as context for own activity. Specifically, Schmidt (2002) argues for awareness to be seen in conjunction with awareness of something, i.e., being or becoming aware of specific aspects of an unfolding situation. Awareness is therefore closely coupled to actual work practice.

Arguably, and in relation to for example the crew on the frigate bridge, awareness implies to be aware or become aware of work practice seen in context, thus to understand what is happening in one's surroundings and to further apply this knowledge into practical use. With this in mind, there are two distinct yet interdependent dimensions of importance. Firstly, professionals need to *monitor* the work of colleagues and secondly, at the same time be able to *display* own work activity. To understand awareness, a particular analytical interest then lies in exploring and identifying how professionals manage simultaneously to monitor and visualize work practice, and the practical strategies and the resources they rely on using.

Klemets and Evjemo (2014) have studied hospital collaborative work among nurses who rely extensively on various technologies to communicate and coordinate departmental work. The authors show how nurses handle (and often struggles) interruptions during ordinary, routine work. Klemets and Evjemo (2014) demonstrate that not all interruptions experienced by nurses via their phones are unwanted, i.e., interruptions are dual in nature. Collaboration and coordination characterizes nurses' work – they are highly dependent on each other – and the principal strategies for jointly managing interruptions is to make own work *visible* and at the same time to *monitor* the work of others to enable collegial support. The applied implications of the study are guidelines to use when designing new functionality in the nurse call system so as to facilitate nurses' awareness and strengthen patient safety.

THE PROPERTIES OF PROSPECTIVE SENSEMAKING

Rosness et al. (2015) show that *prospective sensemaking* is not established as a distinct perspective in the scientific research literature, but there are some publications that in some way use the term. They cover a broad range of topics from technology development, consultancy, decision-making in organizations to patient safety to mention a few. Specifically, Rosness et al. (2015) argue that properties with prospective sensemaking and the relationship to (traditional) retrospective sensemaking is seldom debated in the research literature, and they argue that important aspects with making sense of something is missed out. Rosness et al. (2015) define prospective sensemaking as, "… sensemaking processes where the attention and concern of people is primarily directed at events that may occur in the future" (Rosness et al., 2015), a definition this chapter will use in the upcoming discussion of the AIBN report's accident analysis. Importantly, "prospective" processes do not exclude retrospective processes (Rosness et al., 2015). Drawing on past experience to make sense of possi-ble future events is indeed relevant, just as the above definition implicitly recognizes by the use of "primarily".

According to Weick (1995) expectations are important drivers for sensemaking – the effects of expectations on the interpretation of input from the environment are in particular focus. At the same time, the explicit formation of expectations is of par-ticular interest to understand sensemaking processes, because it involves attention on events that might occur ahead in time, i.e., prospective sensemaking. This chap-ter acknowledges the need to explore how professionals' expectations are formed, in particular how expectations become configured and at the same time are embedded within the local context of operations, i.e., through situated, coordinative practices on a ship's bridge. If one takes the example of the actions of the OOW on *KNM Helge Ingstad's* bridge, the AIBN accident report informs us that the tanker *Sola TS* was mistakenly taken to be the oil refinery located onshore. Arguably, the OOW on *KNM Helge Ingstad's* bridge was unable to construct or recognize a pattern of future events that made the available information (cues) sensible. Consequently, *Sola TS's* incitement to the frigate to immediately change course just prior to the collision was not perceived as rational by the crew on *KNM Helge Ingstad's* bridge. Rosness et al. (2015:60) developed the term from observational studies within surgical hospital settings in Norway. Table 5.1 provides an overview of the six main properties of prospective sensemaking as elaborated by Rosness et al. (2015).

Prospective sensemaking is thus empirically rooted, and a couple of empirical examples from the original study are appropriate. *Time* is an important dimension considering how expectations about future events are formed. Firstly, the follow-ing quote illustrates how the scrub nurse attempts to be one or two steps ahead the surgeon, "… The surgeon is looking for a forceps on the table and then asks for it. He then discovers that the scrub nurse already has the forceps ready in her hand" (Rosness et al., 2015:60). By continually observing the surgeon, the scrub nurse is able to make expectations on what equipment he or she will need next and thus pre-pare in advance. This example is directed at the immediate future; however, expec-tations related to a longer time horizon are also common and are illustrated by the following quote; "… towards the end of the surgical procedure, the anaesthetic nurse explains that she gave a little extra pain reliever intravenously. It will have effect for some time after the patient has awakened from general anaesthesia. She does not fully trust the plexus anaesthesia because the patient has so much fat that it can be difficult to hit the correct nerve" (Rosness et al., 2015:60). Arguably, both nurses' capacity to stay ahead of events that they expect is imperative for safe and efficient conduct of surgical procedures.

A similar argument will apply to the events preceding the collision in Hjeltefjord. Exploring the formation of crew expectations, the strategies and resources applied are particularly intriguing, especially the relationship between expectations of future events as a basis for crew collaboration (action) and how expectations are either reinforced or weakened through real-time access to information as an event evolves. In line with Rosness et al. (2015), this chapter views sensemaking processes as evolving phenomena, i.e., not a static state or end goal in itself. Similarly, the starting point is that prospective sensemaking is a key prerequisite for both safe and efficient maritime traffic.

TABLE 5.1

Six Properties of Prospective Sensemaking Based on Observational and Interview Material with Medical Professionals in Operating Teams

Main Characteristics (Rosness et al., 2015)	Analytical Focus
The persons involved are primarily concerned with their own and the team's successful handling of events in the near or intermediate future, ranging from seconds and minutes to weeks and months into the future. Their attention is thus directed at the future, rather than the past.	Time span, near or intermediate future
Prospective sensemaking does not necessarily require strong external cues or triggering events to occur. Although retrospective sensemaking activities are typically triggered or intensified by uncertainty or ambiguity, prospective sensemaking also occurs spontaneously, as a "natural" part of the work practice.	Naturally occurring work practice
Prospective sensemaking relies on both verbal and non-verbal communication, including observation of the actions of others and of the effects of those actions.	Verbal and non-verbal communicative actions
Prospective sensemaking can be open to the possibility of alternative chains of events—the future may be conceived as an event tree rather than a single path of events. An implication of this is that prospective sensemaking allows for ambiguity and uncertainty.	Events are non-linear – implies uncertainty/complexity
The main outcomes of successful prospective sensemaking are practical preparations to handle possible future events, mental preparedness to interpret future events, and improved coordination in tasks involving intertwined actions of two or more persons.	Outcomes; practical preparations, mental preparedness, coordination between/among several people
The process of prospective sensemaking may involve human as well as non-human actors, including different forms of representations or models.	Human and non-human actors, e.g., technology

Source: Modified, based on Rosness et al. (2015).

OPERATIONALIZATION OF THE PROPERTIES OF PROSPECTIVE SENSEMAKING

Let us return to when *Sola TS* left the Sture terminal for the first example. The AIBN report explains that the traffic controller did not feel that traffic regulation or information was needed for any of the other vessels in the area. The traffic controller's work screen was also zoomed in on *Sola TS* at the time *KNM Helge Ingstad* was not followed on radar, which according to AIBN led the traffic controller to simply forget the frigate. Arguably, the traffic controllers' (Fedje VTS) initial reception of the call from *KNM Helge Ingstad* creates an expectation (or several) regarding future events in which the frigate is involved. However, the call from *Sola TS* causes the traffic controller to zoom in on the Sture terminal and leaves the screen to remain so even after the tanker has left the terminal, which contributes to the traffic controller not having an overview of the frigate's position just before the collision. Thus, there may appear to be a distinction between short-term expectation versus long-term expectation based on the traffic controller's handling of the two ships, which makes it interesting to investigate more closely how a *time dimension* associated with different expectations influences preparation.

It would be interesting to examine further whether the actions relating to the former are carried out at the expense of the latter and whether there is a tension between these two that can be detrimental for the opportunity the traffic controller has to strengthen or weaken own expectations during the course of the event. It also appears (from the perspective of Fedje VTS) that the sailings of the two ships from the beginning are somewhat disconnected events, which arguably can make coordination challenging when the situation gradually tightens.

Moreover, at the same time as *Sola TS* left the Sture terminal, a change of OOW's began on the frigate bridge, which is our second example. According to the AIBN report, both the outgoing and the incoming OOW's relate to an object on starboard side of the frigate. The report describes that the object was identified visually as well as using radar with subsequent discussion between the OOW's related to the object's identity. A positive identification was not carried out, but both officers were of the opinion that the object was stationary and posed no risk to the frigate. The AIBN report states that after the shift, the opinion on the frigate bridge prevailed that *Sola TS* was a stationary object, which was also the starting point for the oncoming officer's subsequent actions. The situation on the frigate bridge was such that the training of a new OOW received a good deal of attention, which affected the crew's capacity to monitor the overall together with specific aspects of the traffic situation in Hjeltefjord.

The accident report also points out that the starboard look was unmanned, which further reduced the possibility of changing the perception of the traffic situation on the bridge. Since the incoming OOW was of the opinion that the *Sola TS* was a stationary object, the tanker was not radar tracked, which also meant that there was no alarm when the ships approached each other. Initially in example two, we saw how the overlapping OOW's discussed what the unidentified object was – the consequence of this interaction being a common anchorage linked to expectations of what will be current events in the future. It can thus be argued that anchoring which events are relevant in the future involves a clear social dimension through *negotiations* between different actors in relation to the interpretation of available information, which may challenge the idea that, for example, functional redundancy will most often minimize the risk of accidents.

Furthermore, the third example involves the use of distinct technology, e.g., that the radio communication between *Sola TS* and Fedje VTS was not heard by *KNM Helge Ingstad*. This information was broadcasted on an open frequency and such information can be imperative to either strengthen or weaken own expectations, and it connects in general with how key actors *configure actions* to prepare so as to be ahead of any upcoming event. The AIBN report speculates that this is related to either the switching of OOW on the frigate bridge or that Fedje VTS did not explicitly inform *KNM Helge Ingstad* or other vessels in the area about *Sola TS'* intentions. Alternatively, the frigate had distinct routines or practices related to listening to radio which caused the communication between *Sola TS* and Fedje VTS at this particular moment in time not to be heard. Moreover, the frigate was not plotted on *Sola TS*'s radar, which the AIBN points out may have to do with little explicit communication between the crew of *Sola TS* and the pilot. According to AIBN, this also affects the understanding of the situation on *Sola TS's* bridge. On the other hand, *KNM Helge Ingstad* was indeed sailing without AIS signals, which resulted in the name not appearing at *Sola TS*. At the same time, the frigate was not followed electronically by Fedje VTS. Technologies were available to all crews which could have made each other's identity and position more visible, but for various reasons the technology was not used as it could. It would be valuable to learn more about such choices –why any given technology's (designed) affordances are not manifested in use involves understanding *how* actors choose to prepare given the events one expects.

Therefore, and based on Table 5.1 and the examples discussed above, three questions are proposed that are particularly relevant for an operationalization of prospective sensemaking to better understand the accident in Hjeltefjord:

1. How are distinct events coupled in time and space in the sociotechnical system in question?
2. How are events made relevant through interaction between key players?
3. In what ways do key players prepare to handle expected events including the principal resources used?

The first question thus addresses whether tension exists between long-term and short-term expectations that hamper checking and or verifying own expectations during course of events. The second question addresses the social construction of events and particularly how events are rendered intelligible through crew negotiations. The third question explores the practices surrounding preparation, more precisely situated action and the embedded (technological) resources used by the crew.

A DELIMITATION OF KEY ANALYTICAL DIMENSIONS FOR THE OPERATIONALIZATION OF PROSPECTIVE SENSEMAKING

Figure 5.1 conceptualizes distinct analytical dimensions coupled to the three derived questions above. The idea is to argue for the need to analytically divide and refine and then see in context specific dimensions that are relevant to understanding as well as evaluate prospective sensemaking, so as to also facilitate an expanded understanding of what the AIBN (2019) accident report, for example, describes as *situational understanding* per se. Of course, although an *event, expectations, ways of preparing*, as well as available *information* are mutually dependent dimensions, it is advantageous to analytically separate the elements from one another. The intersection in Figure 5.1 illustrates interdependence while the smaller arrows analytically illustrate how the various dimensions interact and affect each other and how they can be analysed. An event involves a time-limited situation while expectation is linked to the assumption of a future event. Ways of preparing imply the actual practices coupled to how crew prepare while information/cues are

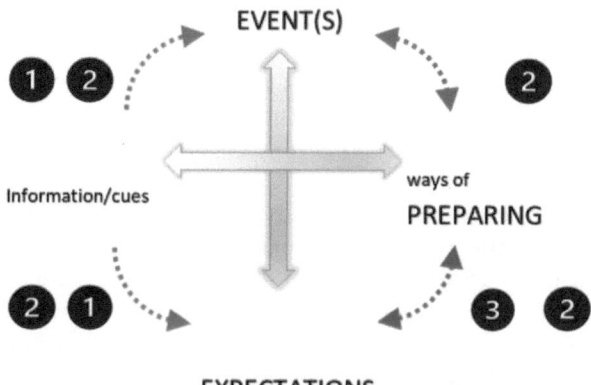

FIGURE 5.1 Analytical (and conceptual) delimitation of prospective sensemaking.

available information resources related to future events. The numbering refers to the previous three questions and the operationalization of prospective sensemaking, more specifically the analytical dimensions they are intended to answer. In example two above, both OOW's on *KNM Helge Ingstad* formed expectations based on what they believed the object was, including what, if any, events would be triggered in the future. The lights from *Sola TS* are examples of information, i.e., a cue that formed expectations but also events. For analytical purposes, information/cues impact events and expectations in a one-way relationship. Moreover, events affect ways of preparing and vice versa. An event can trigger distinct ways of preparing, e.g., the frigate started to turn slightly based on the OOW acknowledging the need to increase distance to the (now moving *Sola TS*) object.

At the same time, it is conceivable that the way one prepares, for example, looking at a radar screen could influence how an event is constituted but also expectations per se related to the event. Important in this context is which *analytical unit* is the starting point for analysis when, for example, the purpose is to explore the properties of prospective sensemaking at the bridge of *KNM Helge Ingstad*.

SOME LEARNING POINTS FROM AN OPERATIONALIZATION OF PROSPECTIVE SENSEMAKING

The learning points from the events prior to the collision in Hjeltefjord in the light of prospective sensemaking are about expanding the understanding of how interaction between the mariners takes place, including the resources used, thus laying the grounds for a more comprehensive systemic understanding of what happened. Consequently, applied prospective sensemaking is also about expanding the traditional understanding of what human factors entail by analytically refining key aspects of key participants' interactions, which the operationalization and conceptualization of prospective sensemaking attempts. By taking the main characteristics of prospective sensemaking as described by Rosness et al. (2015) and also adding analytical focus, the *properties* of prospective sensemaking (Table 5.1) are arguably well suited to explore the *type* of accident (during normal operations) that occurred between *KNM Helge Ingstad* and *Sola TS*. Based on the review of the course of events from the Accident Investigation Board's report (AIBN, 2019), three main questions have been derived where the idea is to elucidate key aspects of prospective sensemaking, which the conceptualization of analytical dimensions (Figure 5.1) thereafter elaborates.

An important element of the applicability associated with prospective sensemaking is about transferability to other situations, as well as to other domains. It can be envisaged that such a perspective is also useful in the investigation in, for example, aviation including the Boeing 737 MAX accidents where weak human-centred design is part of the picture (Endsley, 2019), and to issues of unfortunate human–automation interaction, i.e., the safety paradox (Evjemo & Johnsen, 2019). The transferability of prospective sensemaking to similar domains is thus rooted in the operationalization and conceptualization of the term. Applied use of prospective sensemaking means that one can complement the most often used situational understanding, i.e., situational awareness (Endsley, 1995), however also going beyond a foremost cognitive approach to sensemaking (Weick 1995) via incorporating the social and a *forward*-oriented perspective. At the same time, the operationalization and conceptualization of prospective sensemaking involves a clear analytical focus on technology use as situated (Suchman, 1987; 2007), which is also illustrated by the examples from Hjeltefjord.

There are potentially several learning points from an applied use of prospective sensemaking for future accident investigations that are important to bear in mind. Firstly, there are specific methodological issues. When investigating these types of accidents, it is necessary to identify and map in detail the sequential ordering of action, which applies to both the overall movement of a vessel and to the interaction between crew on the bridge and between the crew and other actors such as in our example, Fedje VTS. Furthermore, a specific analytical focus on the use of participants' resources including the use of distinct technology is imperative to understand why events evolved the way they did. Secondly, it is important to consider how empirical material is collected. What methods are most appropriate to capture both participants' meanings as well as action? Rosness et al. (2015) used a variety of methods in their study of hospital work and Evjemo (2017) argues the need to combine interviews and observational work to ensure sufficient *proximity* to the empirical data in complex work settings where compliance to standards are imperative to safety. Thirdly, and from a practical standpoint, one can also imagine that future formal team safety training including technology use can incorporate ideas from prospective sensemaking, e.g. available interactional resources and their use to strengthen the crews' ability to identify and handle safety critical situations in a more proactive manner.

CONCLUDING REMARKS

It is worth recapitulating the point of departure for the chapter, namely that the purpose was not to assess whether prospective sensemaking was evident among the crew on the frigate and the tanker or within the traffic control centre. The purpose of this chapter was to demonstrate how prospective sensemaking can be operationalized based on how the events prior to the accident are identified in the Accident Investigation Board report, where the root cause was said to be misunderstanding of the situation on the frigate bridge. At the same time, specific criticism is put forward towards both the tanker and the shipping control centre without describing the details of collaboration and technology use including a focus on how key actors plan and organize for future events. This chapter has argued for the relevance of prospective sensemaking as a (future) tool to increase understanding of interaction in complex socio-technical work environments while the chapter has also demonstrated how such applied use of prospective sensemaking can be grounded in an operationalization and conceptualization that is empirically rooted, so as to contribute to safeguarding safety in complex organizational domains.

REFERENCES

AIBN (2019). Delrapport 1 Om Kollisjonen Mellom Fregatten Knm Helge Ingstad Og Tankbåten Sola TS Utenfor Stureterminalen I Hjeltefjorden, Hordaland, 8 November 2018. Statens Havarikommisjon for Transport.

De Boer, R. & Dekker, S. (2017). Models of automation surprise: Results of a field survey in aviation. *Safety* 3: 1–11.

Dekker, S. (2013). On the epistemology and ethics of communicating a cartesian consciousness. *Safety Science* 56: 96–99.

Dourish, P. & Bellotti, V. (2002). Awareness and coordination in shared workspaces, in: *Proceedings of the 1992 ACM Conference on Computer-Supported Cooperative Work.* ACM, Toronto, Ontario, Canada.

Endsley, M. R. (1995). Toward a theory of situation awareness in dynamic systems. *Human Factors* 37: 32–64.

Endsley, M. R. (2019). *Human Factors & Aviation Safety*, Testimony to the United States House of Representatives Hearing on Boeing 737-Max8 Crashes—December 11, 2019.

Evjemo T. E. & Johnsen, S.O. (2019). *Lessons learned from increased automation in aviation: the paradox related to the high degree of safety.* ESREL 2019.

Evjemo, T. E. (2017). Understanding (organizational) practices in safety research: Proximity versus distance and some research implications. In Bernatik, A., Kocurkova, L. & Jørgensen, K. (eds). *Prevention of Accidents at Work.* London: CRC Press.

Klemets, J. & Evjemo, T. E. (2014). Technology-mediated awareness: Facilitating the handling of (un)wanted interruptions in a hospital setting. *International Journal of Medical Informatics* 83(9): 670–682.

Rosness, R., Evjemo, T. E., Haavik, T. & Wærø, I. (2016). Prospective sensemaking in the operating theatre. *Cognition, Technology & Work* 18(1): 53–69.

Schmidt, K. (2002). The problem with 'awareness': Introductory remarks on 'awareness in CSCW'. *Computer Supported Cooperative Work* 11: 285–298.

Suchman, L. (1987). *Plans and Situated Actions.* New York: Cambridge University Press.

Suchman, L. (2007). *Human-Machine Reconfigurations: Plans and Situated Actions.* New York: Cambridge University Press.

Weick, K. E. (1995). *Sensemaking in Organizations.* Thousand Oaks: SAGE.

Weick, K. E., Sutcliffe, K. M. & Obstfeld, D. (2005). Organizing and the process of sensemaking. *Organization Science* 16(4): 409–421.

6 The Challenges of Sensemaking and Human Factors in the Maritime Sector – Exploring the *Helge Ingstad* Accident

S. O. Johnsen
SINTEF

CONTENTS

INTRODUCTION

In this chapter, we have discussed the challenges of sensemaking and human factors (HF) in the maritime sector based on several accident and incident reports. The *Helge Ingstad* collision has been of special interest, as described in the accident report from the Accident Investigation Board in Norway, AIBN (2019). The following summary documents the highlights of the accident and is followed by a more systematic exploration later in the chapter.

The frigate *Helge Ingstad* and the tanker *Sola TS* collided in Hjeltefjorden at 04:01:15 of 8 November 2018. A total of seven persons were present on the bridge of *Helge Ingstad*. The officer of the watch (OOW) was in charge from 03:53, i.e. the responsibility changed short time before the collision. The tanker *Sola TS* had several persons on the bridge, the pilot, the master, the second mate and the helmsman. *Sola TS* had left the Sture terminal at 03:36. The traffic in Hjeltefjorden was under surveillance by the Fedje Vessel Traffic Service (VTS), and three other ships were in the vicinity of *Helge Ingstad* and *Sola TS*, mowing towards *Helge Ingstad*. No personnel fatalities were recorded as a consequence of the collision, but the frigate *Helge Ingstad* sank. The cost of the frigate was initially 4,000 Mill NOK, and the cost of a new frigate is estimated to 11,000 to 13,000 Mill NOK.

This is an accident in a complex environment, with several involved actors that has to collaborate and being dependent on critical control systems such as radar and map systems (Electronic Chart Display and Information System, ECDIS), thus control and a continuous sensemaking process are important. Automation and control systems relieves people of tasks, but automation challenges sensemaking and requires more, not less interaction design, interface design and attention to training (Parasuraman & Riley, 1997). Thus, to understand accidents in an environment of more automation and control systems, we have explored several maritime accident reports with reliance on automation and control systems on the bridge, in order to identify general challenges of sensemaking and HF.

By sensemaking, in this chapter, we build on Kilskar et al. (2020), defining sensemaking as a dynamic iterative process of observing (cues), orienting and acting in a social setting, thereby creating a shared understanding. Sensemaking is influenced by HF. The three major ergonomic areas in the HF discipline are organizational, cognitive and physical, Karwowski (2012). Organizational ergonomics refers to responsibilities, work process, operational philosophies, Crew Resource Management (CRM), etc. Cognitive ergonomics refers to task analysis, workload, interaction design, human–machine interfaces (HMI), alarm philosophies, etc. Finally, physical ergonomics refers to issues relevant to workplace layout, working environment (climate, noise), etc. These areas are the foundation of the science of human factors, as described in Lee et al. (2017).

As in all accidents, the root causes of the accident are a combination of technical, human and organizational issues. We see the accident as a consequence of deeper challenges with the whole system, i.e. combination of issues such as poor design of control systems, poor training, mental overload and fatigue (Dekker, 2005). The focus of this chapter has been the sensemaking among the different actors (*Sola TS* and the VTS) and on the *Helge Ingstad* bridge prior to this accident, how the sensemaking process was influenced by factors from the environment, clues from the systems, and how sensemaking could have been more robust.

We use the term safety critical to denote situations or operations that, if they go wrong, have a large potential for causing harm to people, property or environment. Critical operations on the bridge include voyage planning, navigation, positioning and manoeuvring the ship during the voyage. Key systems used on the bridge are DP systems (dynamic position) and navigation systems (radar and ECDIS).

Outside the maritime domain, there are some useful lessons to be learned from industries with a high focus on safety and reliability. Exploring the Macondo Accident from the oil and gas industry and the Boeing Max accidents from aviation, there are key lessons to be learned related to focus on HF in design and operations to build sensemaking.

The Macondo blowout in 2010 killed 11 workers, released 4.9 million barrels of oil and generated expenditures of more than 61.6 Billion USD (*Washington Post* 15th of June 2016). In the accident report CSB (2016), HF were highlighted, i.e. "Industry's focus must shift from correcting individual 'errors' identified post-incident to a systematic approach for managing human factors." Furthermore, the report highlights that "the lack of effective integration of human factors into the design, planning, and execution of drilling and completions activities ... and it illustrates a demonstrable gap in US offshore regulation and guidance to incorporate more robust management of human factors." Specific areas mentioned were the importance of HF engineering in design of safety-critical systems; the need for focus on non-technical skills (such as communication, teamwork and decision-making between different actors) and assessment of safety-critical tasks and identification of controls that could maximize the likelihood of successful human performance through improved sensemaking. (HF design activity includes the design of HMI as a part of cognitive ergonomics.)

The Boeing Max 737 was grounded worldwide in March 2019, after 346 people died in two crashes, i.e. Lion Air Flight 610 on October 29, 2018, and Ethiopian Airlines Flight 302 on March 10, 2019. The Boeing Max 737 disasters were caused by many factors, among others complex control systems, failures of sensors, challenges of sensemaking during critical operations and poor design. In a hearing to the US Congress, Endsley (2019) pointed out that development of the critical control systems should be in compliance with HF design standards (ensuring that design should support cognitive ergonomics); professionals trained in HF Engineering should be included on the design team and robust human user testing should be conducted to validate system design.

RESEARCH QUESTION

Key issues to be explored in this chapter has been the sensemaking process among the different actors involved in these accidents and how the sensemaking process could have been improved, i.e. the two research questions are:

What are the key issues influencing the sensemaking process during operations of safety-critical operations?
What are the key issues influencing the sensemaking process in design of control systems used in safety-critical operations?

METHOD AND APPROACH

We have based our approach on review of the accident reports from *Helge Ingstad*, supplied with interviews with actors involved in the accident analysis in addition to a review of 19 accident reports focusing on control systems on the bridge.

To help analyse the accidents, we have structured our review based on the CRIOP method, Sintef (2011). CRIOP is an internationally accepted method that is being used as a good practice guideline when designing control facilities, Aas et al. (2009). CRIOP is focusing on HF supporting sensemaking when using control systems. Key issues from CRIOP that has been explored are:

Safety-oriented design based on a task analysis. Exploring on how tasks must be designed through a balance between automated operations and human operated tasks supported by HMI, i.e. conditions supporting cognitive ergonomics. Support from regulations to perform necessary safety-oriented design.

Design of control and safety systems to support sensemaking and cues, design of HMI and support of sensemaking, supporting cognitive ergonomics through high-performance HMI and clues that can support the operators.

Alarms, design and handling of alarms to understand and handle critical situations – avoiding undue cognitive workload.

Job organization and planning, based on safety-critical tasks, cognitive and physical workload (describing responsibilities of operations and high-level work procedures) planning, i.e. safety management.

Procedures and work descriptions, based on task analysis and established together with the users.

Physical layout of work place and working environment, based on systematic task analysis and how jobs are organized, supporting all tasks (especially safety critical).

Competence and training of the involved actors in their different roles using control and safety systems and appropriate procedures.

We have performed a review of 19 maritime accident reports related to control systems on the bridge. (To focus our review on systems used on the bridge, we have selected accidents that involved onboard control systems, i.e. accidents which involved onboard electronics/control systems in some shape or form.)

We have tried to structure root causes based on the above taxonomy from CRIOP, i.e. loss of situational awareness (i.e. poor sensemaking process), poor cognitive ergonomics design (poor redundancy, poor cues), poor planning, poor work load assessment, alarm issues, poor competence and training, poor safety management, poor support from regulation and poor ergonomic layout,.

The 19 accident reports were selected in collaboration with an expert within the area of maritime safety. The review included 14 Marine Accident Investigation Branch (MAIB) investigation reports from accident occurring in the period 2005–2016 as well as 5 other investigation reports from accidents occurring in the period 1995–2008.

ANALYSIS OF CONTROL SYSTEMS ON THE BRIDGE

In the following, we have summarized the experiences from our review of 19 maritime accident reports related to control systems on the bridge. The issues are described in more detail in Johnsen et al. (2019). A key statement from the chief inspector of MAIB in one of the reports was: "this is the third grounding investigated by the MAIB where watchkeepers' failure to use ECDIS properly has been identified as one of the causal factors. In 2014 there are over 30 manufacturers of ECDIS equipment, each with their own designs of user interface, and little evidence that a common approach is developing," Johnsen et al. (2019).

Loss of situational awareness (in ten accident reports) due to poor monitoring of position; distraction due to workload; unsafe navigation practices; poor passage planning; insufficient understanding of control system; misinterpretation of the nature of malfunctions.

Poor redundancy/alternatives (in nine accident reports) in terms of poor orga-nizational redundancy of coastguard; poor backup of equipment; no contingency planning; poor route planning not cross-checked; undue reliance on the ECDIS; practice of operating with watertight doors open; not using at least two independent sources to verify position; no installed navigation autopilot with alarm when discrep-ancies were detected. In summary, there is a need to establish resilience in critical operations such as planning and navigation.

Alarm-related issues (in ten accident reports) including disabling of alarms and thus removing necessary barriers of imminent danger; alarm system silenced, missing entering of passage plan; ECDIS not utilized effectively as navigation aid and audible alarm disabled; ECDIS safety setting not appropriate – audible alarm inoperative – and defect of alarm system not being reported; system giving alarm per minute and overwhelming the watchkeeper; poor understanding of the system and relationship of alarms; navigation equipment ineffective and not set-up to use all safety features; no installation of alarm comparing position from multiple indepen-dent positions. In summary, alarm design is a key issue.

Insufficient training (in 15 accident reports) in terms of no emergency pre-paredness training; operator not qualified and not supervised; untrained in the use of the ECS and unaware of user support; no training in use of ECDIS and no safety procedures established; marked differences in ECDIS systems such as menus, termi-nology and interfaces; poor training of electronic support systems/main engine con-trol systems; poor training in use of the integrated navigation system; poor training in crew resource management and emergency communication; poor focus on con-tinuous professional development and skill retainment. In summary, poor training seems to be a key issue in many accidents using electronic systems (ECDIS, voyage management system, etc.). The accident reports raise the issue of usability and user involvement from design through acceptance of these electronic systems – are the systems so poorly made that they are a challenge to use?

Lacking or insufficient passage planning (in eight accident reports) including poor passage planning and poor checking and approval of the route (i.e. grounding was inevitable due to vessel draught and depth of water); poor utilization of ECS or ECDIS for passage planning – the system would have given alarms early. In sum-mary, the quality of planning is poor and the support from the ECDIS is often miss-ing (either due to poor training or poor design).

Poor or missing work load assessment (in eight accident reports), the sole bridge watch-keeper having to undertake passage planning and chart corrections and bridge manning was insufficient; the coastguard being distracted and did not send warning due to chronic manpower shortages; the bridge team having to provide administrative information when they should focus on safety of vessel passage; the bridge missing an appropriately certified third person; a widespread deselection of automated functions in ECDIS to reduce workload (indicative of wider problems with the ECDIS design). In summary, organizational factors as well as design issues contribute to work load and fatigue.

Poor (safety) management (in 14 accident reports) including the harbour not having a risk assessment or safety management plan in the pilotage area; the crew seeing no value in safety management; the master providing insufficient safety focus/culture; poor clarity in responsibility during watch; inefficient safety audits based on ISM code; poor risk assessment prior to work on ballasting; poor passage planning – not cross-checked – and mitigating actions not performed. In summary, risk-based focus of operations is sometimes missing, and there is variability.

Poor system design or display layout (in eight accident reports) in terms of deficiencies in design and implementation of the integrated bridge system and in the procedures for its operation; widespread de-selection of automated functions in ECDIS that is indicative of wider problems with ECDIS; ECDIS not used as expected by the regulators or equipment manufacturer; ECDIS safeguards intended to prevent grounding were overlooked, disabled or ignored; MAIB chief inspector said: "this is the third grounding investigated by the MAIB where watchkeepers' failure to use ECDIS properly has been identified as one of the causal factors. In 2014 there are over 30 manufacturers of ECDIS equipment, each with their own designs of user inter-face, and little evidence that a common approach is developing." In summary, there is a need for standardization, improved user-based design and user-based acceptance testing in normal operations and during critical operations.

Key Findings

The ability to understand the status at a glance (and get an understanding of key risks) is missing in some of the bridge systems. When focusing on sensemaking, the usability qualities of the control systems (ECDIS, Bridge systems, DPS) are poor and should improve. The poor usability also influences the needed training regime, since training and competence development seems challenging (due to system complexity) and sometimes missing (due to costs, poor practice and missing regulation of training). Passage planning seems poor due to poor usability and missing operational procedures – thus the systems do not support sensemaking as they should. The alarm systems have not been adapted to the users' workload and system understanding, thus alarms seem a disturbance and not an input to improved sensemaking.

Design of organizational procedures and work should be performed together with the seafarers to ensure usability of procedures and checklists; and clarity in responsibility and proper work load. Too high work load may lead to stress and challenges sensemaking and understanding.

Designing of alarms should be performed to ensure that alarms are designed to support sensemaking and not stress the operators with too many alarms, i.e. more than six alarms each hour as specified by EEMUA-191 (2013).

There is a need to increase focus on user-centric design principles. Poor design is a significant contributor to maritime accidents. There is a need for improved regulations and standards related to use of integrated bridge systems and ECDIS. Design of bridges and control systems should be based on user-centric design principles, involvement from HF experts and should be subject of inspections, regulators and workforce attention. Benefits of user-centric design should be highlighted through research.

The quality of alarm systems is poor, and alarm guidelines and standards should be established based on industry best practices considering human limitations. An integrated alarm philosophy must be established for all systems on the bridge – not individually for each system. Collaboration between regulators, industry and classification society should be prioritized to speed up adaption.

KEY ISSUES OF THE *HELGE INGSTAD* ACCIDENT

We have explored the *Helge Ingstad* collision based on key elements in the accident report based on the methods as described earlier. Main points of the accident were: The frigate *Helge Ingstad* and the tanker *Sola TS* collided in Hjeltefjorden at 04:01:15. The OOW was in charge from 03:53. The tanker *Sola* had left the Sture terminal at 03:36. The traffic in Hjeltefjorden were under surveillance by the Fedje VTS. The OOW did not identify *Sola TS* as a ship close by, but had a perception that what he could observe were a lighted part of the Sture terminal.

We have highlighted the following areas from the accident report:

Positive focus on a broad system perspective, not blaming human errors
Positive focus on sensemaking and situational awareness in the report
Poor analysis and piecemeal design of the bridge – poor task analysis and assessment of mental workload, and poor design of the totality of the workplace decreasing sensemaking possibility; failure to integrate VHF with radar and ECDIS equipment
Poor design of work environment – not sufficient focus on noise
Poor design of workload – missing alarm philosophy
Poor operational selection of crew – only 3 of 7 had no visual impairment
Poor operational decisions at margins – working at margins with high mental workload (both training and safety-critical operations late night – with fatigue/slower perceptions)
Breakdown in sensemaking at the *Helge Ingstad* Bridge and between VTS, *Helge Ingstad* and *TS Sola*

Broad system perspective – The accident report from the AIBN (2019) presented an impressive collection of technical, organizational and HF issues, with a focus on trying to describe and understand the accident, instead of blaming human actors (that often perform within a demanding framework that is established prior to the accident). The accident report presents a sequence of what happened during the accident and a careful analysis of framework conditions and background, with much more detailed findings and analysis than recent accident reports from similar maritime accidents such as USS John S. McCain (2017) and USS Fitzgerald (2017). The accident report was careful not to blame individuals but had a system perspective. Earlier, there had been different practices. As an example, in 2000, two accident reports were presented at the same time in Norway: one from the MS Sleipner (1999) maritime accident with a responsible captain in charge of the ship and the other from the Åstad train accident (2000) with a responsible train driver in charge. The Åstad accident report had a system perspective (the train driver was not blamed); the Sleipner accident report blamed the captain and did not focus sufficiently on framework conditions that could be seen as root causes.

Focus on sensemaking – In the *Helge Ingstad* report, the AIBN explored the concept of situational awareness as describe by Endsley (2016), team collaboration and the concept of sensemaking. This framework and understanding of team actions and collaboration must be in place before further causes can be explored, such as technology and organizational issues.

Piecemeal design – The design of command facilities on the bridge should be based on a systematic task analysis of all necessary tasks. The task analysis is used as a basis for responsibilities, designing interfaces and layout decisions, i.e. to understand how equipment such as radar, ECDIS and VHF radio should be placed to support the tasks. The AIBN remarked that the design of the bridge was not optimal to ensure common situational awareness on the bridge. As an example of the missing focus on human factors, the installation of the VHF can be mentioned. The VHF radio had been installed at a later stage, via separate piecemeal installation procedure. At *Helge Ingstad*, the VHF radio was placed in a corner, making it difficult to get information from the radar and ECDIS when using the VHF. In the minutes before the collision, the OOW was using the VHF and talking to *Sola TS* and had not easy access to the radar and ECDIS. When the VHF was installed at one of the other ships, an experienced officer was on the watch, and based on his awareness of critical tasks he managed to get the VHF placed besides the radar and the ECDIS, in order to get an overview of the radar and ECDIS at the same time as communication via the VHF took place. The task analysis is also used as a basis for manning and an assessment of workload in all safety-critical operations. The manning of the bridge had been influenced by a high-level strategy called LMC (lean manning concept), implemented primarily to reduce costs. There was no workload assessment of performing training at the same time as performing safety-critical tasks such as navigation. The design was not based on task analysis (i.e. good design practices), had not performed systematic task analysis of new equipment when it was placed on the bridge and operations were not based on appropriate work load analysis when performing safety-critical tasks. The quality and usability of the ECDIS system seems poor, based on a review of prior accidents. Data from ECDIS indicate that *TS Sola* updated its status from "at quay" to "under way" 30 minutes after actual cast-off, a status information that could have impacted the sensemaking of the OOW. The poor usability, quality and support from the systems (radar and ECDIS) during this accident support the prior accident analysis. There were some adaptations in practices and equipment on the bridge that were mentioned in the accident reports, such as changes in routes and adaptations of night vision. After implementation of the ECDIS system, the ships are more often using the direct route plotted into the ECDIS system; thus they are no longer using the traditional sea routes, close to land on starboard side that ensured that ships were separated when going in the fjord. One adaptation mentioned that some of the lights on the bridge had been taped over in order to preserve "the night vision" of the officers on the watch., i.e. indicting poor consideration of user needs during night time.

Working environment – Noise: The ergonomics of the working environment is the key to ensure that communication supporting sensemaking can take place. Working environment is influenced by many factors such as noise, lightning level, and temperature. Team communication and teamwork on the bridge is dependent on coordinating mechanisms such as shared mental modes, closed-loop communication and mutual trust (Salas et al., 2005). To ensure closed-loop communication, the noise level should be assessed. In the accident report, it is pointed out that "Bridge ventilation system is so noisy that it is difficult for the bridge team to communicate in a normal manner. Excessive levels of noise interfering with voice communication, causing fatigue and degrading overall system reliability, shall be avoided. (noted during visit on-board)." In Sunde et al. (2015), they pointed out that "*All vessel classes, except the coast guard vessels, had noise levels exceeding the* Royal Norwegian Navy (RNoN) *standard's recommended maximum noise levels.*" The background noise level should be below 45 db when performing safety-critical tasks (ref SINTEF, 2011), but the noise level was higher in *Helge Ingstad*, impacting sensemaking.

Workload – alarm philosophy: Mental workload is dependent on alarms and alarm handling. No alarms were given related to the impending collision with *Sola TS*. However, in the accident report, it was mentioned that the bridge handled 12 alarms in the last 14 minutes; this is almost one alarm each minute. (These were alarms related to objects that the bridge were aware of and thought they could control, no alarms from *Sola TS* that they collided with.) This is a fairly high mental workload, even if it seems a controllable environment. The international recognized alarm standard EEMUA-191 (2013) specifies that the maximum number of important alarms that can be handled is 6 alarms in 1 hour, i.e. 10 minutes between each alarm. Thus, the alarm philosophy should have been discussed further.

Poor operational selection of crew – 3 of 7 had no visual impairments. In the AIBN report, it was mentioned that 3 of the 7 persons on the bridge had no visual impairment, i.e. some of the crew on the bridge had some sort of visual impairment. It is uncertain if this impacted the accident, but team cognition could have been impacted, and these facts should have been known when responsibilities were planned on the bridge.

Poor design impacting operational decisions at margins – The *Helge Ingstad* performed fairly critical operations at night time, where resources were allocated to training and navigating in an area with several ships. This environment creates fairly high mental workload, and it is know that many disasters happen during night time, i.e. this is a period of operational risks. A systematic task analysis, analysis of mental workload and operational risk analysis should have been a part of design of the manning, the design of procedures to enable safe operations during the training task. These procedures were missing and are due to missing design of operational procedures based on a systematic task analysis.

Breakdown of sensemaking between key actors during the collision – If we look on the extended team of key actors (involving *Helge Ingstad*, the VTS and *TS Sola*), the mental models were not shared (i.e. common perception among the actors about position and course) and the closed-loop communication during the emergency was poor (i.e. clarity of responsibilities, communication procedures during an emergency and understanding of who was speaking).

- The VTS got a call from *Helge Ingstad* at 02:38 informing the VTS that they would enter the area of the VTS. At 2:50 *Helge Ingstad* entered the area, but the VTS did not plot *Helge Ingstad* on its radar. At 3:45, the VTS acknowledged that *Sola TS* departed. Three ships going north and two ships (*Helge Ingstad* and *Dr No*) going south – in addition to *Sola TS*. *Helge Ingstad* was poorly identified as it did not use the automatic information system (AIS). At 3:59, the VTS discovered a possible collision vector between *Sola TS* and *Helge Ingstad*. The VTS communicated to *Sola TS* informing at 03:59 that it could be *Helge Ingstad*. At 04:00:44 the VTS told *Helge Ingstad* that they had to do something – but no commands from the VTS until the collision at 04:01:15. The VTS did not understand the situational assessment of *Helge Ingstad* and did not intervene through emergency procedures.

- *Sola TS* was at the brightly lighted Sture terminal and left at 03:36. The pilot at the bridge informed the VTS about their departure at 03:45. ECDIS information was updated later and was not communicated to *Helge Ingstad* at departure, giving latency and a perception that the ship was at the Sture terminal. At 03:52, the bridge saw the ship *Helge Ingstad*. At 3:58, *Sola TS* asked the VTS about *Helge Ingstad*. From 03:59:56, there was communication between *Sola TS* and the OOW. *Sola TS* tried to communicate an impending collision at 04.01.15.

- The responsible OOW at *Helge Ingstad* changed at 03:53 (after being briefed from 03:45 to 03:53). *Sola TS* was visually observed, but as a part of the land-based Sture terminal and not as a ship by most of the attending personnel on the bridge, however, this perception and understanding were not communicated and shared. Verbal clues and information sharing were a challenge due to the noise level. At 04:00, the OOW used the VHF radio away from ECDIS/radar, thus visual clues could not be shared. The OOW had a perception that *Sola TS* was the Sture terminal. Large part of the ship (200 m) was not lighted – it was difficult to see *Sola TS*.

Exploring the accidents in hindsight, we see poor common mental models, poor emergency procedures but also a set of missing redundancies of cues to help the crew and the OOW to identify *Sola TS* as a ship.

CONCLUSIONS AND FURTHER WORK

The *Helge Ingstad* accident should not come as a surprise when looking at the poor quality of cognitive ergonomics on the bridge (i.e. poor task analysis, poor configuration of equipment, poor alarms, poor redundancy of cues). The accident clearly shows the consequences of performing piecemeal building of control systems instead of focusing on a unified sensemaking design based on cognitive ergonomics. The *Helge Ingstad* accident is in line with the analysis we have performed of the 19 other accidents where control systems were involved, with poor human-based design, poor focus on cognitive ergonomics and poor sensemaking in critical operations.

In many ways the crew of *Helge Ingstad* met a combination of many factors that reduced the sensemaking capability of the team, such as:

- Time – 4:01 in night 8 minutes after responsibility was changed on the watch; a time where accidents usually may happen.

- High mental workload and high level of disturbances with 12 alarms in the last 14 minutes (Standards specify 1 important alarm in 10 minutes).: in addition to four moving boats (three going north), while the closest object *Sola TS* (unidentified), where the front of the 200 m long boat was in darkness.
- Poor design of the totality of control systems: the OOW could not see their own position while communicating via the VHF. The VHF radio was placed so that the responsible personnel could not see ECDIS chart and radar at the same time – not accepted as a good design by other ships.
- Poor routines in *Sola TS* that did not record the cast-off on the ECDIS system, giving wrong information to *Helge Ingstad*.
- Poor quality of the alarm systems and alarm philosophy at *Helge Ingstad*.
- Questions continue about high level of background noise at the bridge
- Overview of the total situation and emergency action from the VTS could have been more proactive, in order to control the development. A total overview as interpreted from the VTS could have been shared among the actors, in addition to emergency intervention to halt the accident.

Control systems in shipping are mostly developed based on a technology drive and seldom based on strong involvement from the users. There is an abundance of specific technical standards but poor focus on an unified approach focusing on cognitive issues and sensemaking during safety-critical operations. The industry should learn from good practices in the oil and gas industry and the aviation industry, and also look towards recent accidents in oil and gas industry and the aviation industry to see what happens when sensemaking is ignored. Technology-driven implementation may not support safety of critical tasks. When the users are subjected to stress and complex situations, poor cognitive ergonomics does led to poor sensemaking and conditions leading to accidents. Our suggestion for the future is to prioritize:

- Sensemaking among all the involved actors, through placing responsibility of the total traffic picture on the VTS centrals (so that they can display an overview of traffic giving the ships "overview-at-a-glance").
- Sensemaking on the bridge through a unified bridge system where cognitive ergonomics is adapted to the critical operations being performed. As a part of this, focus on alarm design and standards so that alarms are set and really used within human limitations (i.e. max six alarms in an hour).
- Verification and validation of design and critical operations through exploration of "safety-cases" and verification of training periodically.

REFERENCES

Aas, A. L., Johnsen, S. O., Skramstad, T. (2009). CRIOP: A human factors verification and validation methodology that works in an industrial setting. *Lecture Notes in Computer Science* 5775, 243–256.

AIBN (2019). Part one report on the collision on 8 November 2018 between the frigate HNOMS Helge Ingstad and the oil tanker Sola TS outside the Sture terminal in the Hjeltefjord in Hordaland County.

Åsta accident (2000) from https://en.wikipedia.org/wiki/Åsta_accident

CSB (2016). *U.S. chemical safety and hazard investigation board: Investigation report volume 3*; Report no. 2010-10-i-os 4/12/2016 Drilling rig explosion and fire at the Macondo well.

Dekker, S. (2005). *Ten Questions about Human Error: A New View of Human Factors and System Safety*. Mahwah, NJ: Lawrence Erlbaum.

EEMUA 191 (2013) *Alarm Systems—A Guide to Design, Management and Procurement*. Edition 3, Engineering Equipment and Materials Users Association, London.

Endsley, M. (2019) "Human Factors & Aviation Safety" Testimony to the United States House of Representatives Hearing on Boeing 737-Max8 Crashes—December 11, 2019

Endsley, M. R. (2016). *Designing for Situation Awareness: An Approach to User-Centered Design*. CRC Press, Boca Raton, FL.

Johnsen, S. O., Kilskar, S. S., & Danielsen, B. E. (2019). Improvements in rules and regulations to support sensemaking in safety-critical maritime operations. In *Proceedings of the 29th European Safety and Reliability Conference (ESREL)*. 22–26 September 2019 Hannover, Germany. ESREL 2019.

Karwowski, W. (2012). The discipline of human factors and ergonomics. In Salvendy, G. (Ed.), *Handbook of Human Factors and Ergonomics*. John Wiley and Sons, New York.

Kilskar, S. S., Danielsen, B. E., & Johnsen, S. O. (2020). Sensemaking in critical situations and in relation to resilience—A review. *ASCE-ASME Journal of Risk and Uncertainty in Engineering Systems, Part B: Mechanical Engineering*, 6(1).

Lee, J. D., Wickens, C. D., Liu, Y., & Boyle, L. N. (2017). *Designing for people: An introduction to human factors engineering*. CreateSpace.

MS Sleipner (1999) accident from https://en.wikipedia.org/wiki/MS_Sleipner

Parasuraman, R., & Riley, V. (1997). Humans and automation: Use, misuse, disuse, abuse. *Human Factors*, 39(2), 230–253.

SINTEF (2011). "CRIOP: A scenario method for crisis intervention and operability analysis." SINTEF Report SINTEF A4312.

Salas, E., Sims, D. E., & Burke, C. S. (2005). Is there a "big five" in teamwork? *Small Group Research*, 36(5), 555–599.

Sunde, E., Irgens-Hansen, K., Moen, B. E., Gjestland, T., Koefoed, V. F., Oftedal, G., & Bråtveit, M. (2015). Noise and exposure of personnel aboard vessels in the Royal Norwegian Navy. *Annals of Occupational Hygiene*, 59(2), 182–199.

USS John S. McCain (2017) "USS John S. McCain and Alnic MC collision" see https://en.wikipedia.org/wiki/USS_John_S._McCain_and_Alnic_MC_collision

USS Fitzgerald (2017) "USS Fitzgerald and MV ACX Crystal collision" see https://en.wikipedia.org/wiki/USS_Fitzgerald_and_MV_ACX_Crystal_collision

7 Addressing Human Factors in Ship Design
Shall We?

V. Rumawas
Bærekraftig Arbeidsmiljø WE Sustain

CONTENTS

INTRODUCTION

This chapter will discuss how human factors are addressed in ship design, their implementations and challenges. Research suggests that more than 80% of accidents and incidents at sea are influenced by human errors and human-related factors (Baker & Seah, 2004; International Maritime Organization, 2012; McCafferty & Baker, 2006). The contemporary view is that human error is a consequence of deeper issues with the system, combination of issues such as poor design, poor training, mental overload and fatigue (Dekker, 2005). Issues of design flaws have been reported for quite a while, and it is implied that less adequate design is one significant contributor that instigates human errors (Miller, 1999; Reason, 2000). There is an indication of a gap between existing knowledge (standards, criteria and requirements) and existing ship designs (Rumawas & Asbjørnslett, 2010a, 2011a). Some lessons seemed to be too slowly learned in the maritime industry (Grundevik, Lundh, & Wagner, 2009; Lützhöft & Dekker, 2002; Rumawas & Asbjørnslett, 2014b) as similar issues keep happening.

"Human factors" is a relatively novel concept to naval architects and marine engineers. It covers a broad discipline with various dimensions, including habitability, workability, controllability, occupational health and safety (OHS), maintainability, maneuverability and survivability (Lloyd's Register, 2008, 2009). From the human factors perspective, a ship can be considered as a living space, as well as a working place. A ship is usually equipped with one main control center on the bridge, one local control center for the engine room and more local control centers for other purposes, such as cargo handling and other missions. There are many demanding issues for the crew working on ships, such as fatigue, poor automation, flooding of alarms and imperfect design, leading to systems that are difficult to use (and has to be adapted).

REGULATION

Like other products or facilities in the world, a ship is designed and constructed according to a set of regulations: some are compulsory and some are voluntary. But, unlike most other products or facilities which usually follow regulations applied in the country where they are located or marketed, a ship can be designed and built according to regulations from a country with more lenient protocols. This practice is called flag of convenience. Essentially, there are two sets of regulations that a ship must follow: international standards set by the International Maritime Organization (IMO) and the flag state where the ship is registered. Ship-owners can decide which flag their ships will be registered under. There is another set of standards that ship-owners shall decide for their ship, i.e. those published by the classification societies. The latter is considered more important and significant to ship design as this chapter will present. It is not endorsed by any authority per se, but more for insurance, assurance and marketing purposes.

IMO is a specialized agency of the United Nations which is responsible for the safety and security of shipping and the prevention of marine and atmospheric pollution by ships (International Maritime Organization, 2020). The main role of IMO is to create a regulatory framework for the shipping industry that is fair and effective, universally adopted and universally implemented.

IMO publishes documents in different format: agreements, circulars, codes, conventions, guidelines, manuals, model courses, procedures, recommendations, regulations, rules and resolutions. The key regulatory concepts are:

- Conventions (formal agreements between states – that can become a law)
  - Code: Detailed description of requirements, part of the convention, providing the international standards
- Resolutions (changes to the conventions)
  - Recommendations are guidelines not formally adopted
- Circulars (clarifications, interpretations of codes and conventions)

Several conventions are significantly important for the human factors' perspective:

- COLREGS – International Regulations for Preventing Collisions at Sea
- Load Lines, 1966 – International Convention on Load Lines

- SOLAS – International Convention for the Safety of Life at Sea
- STCW – International Convention on Standards of Training, Certification and Watchkeeping for Seafarers

SHIP DESIGN

Ship design is a complex and multifaceted process, influenced by conventions (regulations), requirements and several actors. A successfully designed ship is the result of close and good cooperation between the designer, the customer, the yard and the equipment suppliers (Vossen, Kleppe, & Hjørungnes, 2013). The process of designing a ship is often represented by a spiral diagram which denotes the sequential and iterative aspects of the process that include: conceptual design, preliminary design, contract design and detailed design (Gale, 2003).

The design process consists of developing requirements, conducting analyses, developing drawings, building electronic models and writing specifications (Ross, 2009).

Traditionally, users' involvement was rather limited in the process of designing a ship. Often the seafarers inherited poor design that they under any circumstances ever been consulted about.

SHIP DESIGN – EXAMPLES OF REQUIREMENTS, CRITERIA AND CLASS NOTATIONS

Ships are designed and constructed according to certain criteria. There are three different sources from which these criteria can be obtained and applied as the acceptability of a vessel: classification society rules, regulatory requirements and ship-owner's requirements (Ashe & Lantz, 2003).

A key actor is the classification society. A classification society is an organization that establishes and applies technical standards in relation to the design, construction and survey of marine-related facilities including ships and offshore structures (IACS, 2004). A vessel that was designed and built according to a set of rules published by a classification society may apply for a Certificate of Classification. Before the certificate is issued, the classification society will perform a survey on the ship to ensure that the rules are followed. Usually ship-owners will pick a collection of class notations when ordering for new ships: for example, ✠1A1, **Supply Vessel, COMF-V(3), Clean, E0, DYNPOS-AUTR, NAUT OSV (A) and OILREC** (Det Norske Veritas, 2009a). The construction symbol ✠ is assigned to a ship built under the supervision of the society. **1A1** is assigned to a ship with hull, machinery, systems and equipment that are in compliance with the standards. **Supply Vessel** notation refers to a ship designed especially for supply services to offshore installations in the North Sea. **COMF-V(3)** refers to a ship that is designed and built with comfort class, covering requirements for noise and vibration at the most basic level. **Clean** notation indicates that the ship is designed with specific requirements for controlling and limiting operational emissions and discharges. **E0** implies that the instrumentation and automation installed to allow for unattended machinery space. **DYNPOS-AUTR** shows that the ship is equipped with a dynamic positioning (DP) system with redundancy in technical design and with an independent joystick system back up, also known as DP system level 2. **NAUT-OSV (A)** denotes that the bridge has been designed in accordance with established functional requirements and principles of ergonomics for reduced workload and improved operational conditions in All Waters (A), including areas with harsh operational and environmental conditions such as the North Sea. Furthermore, the bridge arrangement provides the information and equipment required for safe performance of the functions to be carried out at dedicated workstations and government statutory requirements. Some of these referrals are compulsory and some are voluntary. In most cases, addressing human factors in ship design is optional. **OILREC** indicates that the ship is designed for recovered oil reception and transportation after a spill of oil in emergency situations. The first two notations explained above (**✠1A1**) are compulsory while the rest are optional.

SHIP DESIGN – CLASSIFICATION SOCIETIES

Today there are approximately 50 classification societies in the world. Twelve of them are members of the International Association of Classifications Societies (IACS): American Bureau of Shipping (ABS), Bureau Veritas (BV), China Classification Society (CCS), Croatian Register of Shipping (CRS), Det Norske Veritas – Germanischer Lloyd (DNV GL), Indian Register of Shipping (IRS), Korean Register of Shipping, Lloyd's Register (LR), Nippon Kaiji Kyokai (NK), Polish Register of Shipping (PRS), Registro Italiano Navale (RINA) and Russian Maritime Register of Shipping (RS). Ship-owners are free to choose which classification society to be used to certify their ships.

HUMAN FACTORS IN SHIP DESIGN

In the ship design and construction textbook published by the Society of Naval Architects and Marine Engineers (SNAME) (Calhoun & Stevens, 2003), human factors is defined as a comprehensive term that covers all biomedical and psychological considerations applying to the humans in the system. Human factors is also stated to cover human engineering and life support, personnel selection, training and training equipment, job performance aids, and performance measures and evaluation as well.

LR (Lloyd's Register, 2008) describes human factors as, 'something that concerned with the task people do and the environment they do it in fitting the job to the person'. Human factors considerations in marine design can be broken down into eight dimensions (Lloyd's Register, 2009):

- Habitability: to ensure accommodation, washing and toilet facilities, messrooms, group meeting and exercise areas are comfortable, clean (or cleanable) and convivial
- Maneuverability: to ensure ships have the most appropriate maneuvering capabilities
- Workability: to ensure ships and systems are appropriate for the work situation (context of use)
- Maintainability: to ensure operational maintenance tasks, manuals, diagnostics and schematics are rapid, safe and effective to allow equipment and systems to achieve a specified level of performance

- Controllability: to ensure appropriate integration of people with equipment, systems and interfaces
- Survivability: to ensure that there are adequate firefighting, damage control, lifesaving and security facilities to ensure the safety and security of crew, visitors and passengers
- Occupational health and safety (OHS): to ensure appropriate consideration of the effect of work, the working environment and living conditions on the health, safety and well-being of workers
- System safety: to ensure appropriate consideration of the risks from people using (or misusing) ship systems.

The IMO uses the term "human element" (International Maritime Organization, 2004) which is defined as follows:

> The human element is a complex multi-dimensional issue that affects maritime safety, security and marine environmental protection. It involves the entire spectrum of human activities performed by ships' crews, shore-based management, regulatory bodies, recognized organizations, shipyards, legislators, and other relevant parties, all of whom need to co-operate to address human element issues effectively.

RULES AND REQUIREMENTS RELATED TO HUMAN FACTORS

There are abundant documents published by Classification Societies that cover human factors-related issues (Rumawas & Asbjørnslett, 2010a, 2014a). Table 7.1 shows a list of rules, guides and notes published by ABS. ABS is a Classification Society that consistently show great interest in the issue of human factors.

IMO codes related to human factors are:

- Code on Alerts and Indicators, 2009
- FSS Code – Fire Safety System
- LSA Code – International Life-Saving Appliance Code
- Noise Levels – Code on Noise Levels on Board Ships

In Table 7.2, examples of other IMO publications that are relevant to human factors in ship design are listed.

Thorough investigations (Rumawas & Asbjørnslett, 2010a, 2014a) showed that there are many documents covering human factors issues in ship design. As an example, all human factors dimensions as LR describes are included. They are published with different degrees of enforcement: some are prescriptive and some are obligatory. However, in general, addressing human factors principles in ship design is still optional, unless those related to safety.

System safety is the most frequently mentioned while habitability or comfort is the most extensively covered. Controllability such as design of alarms and workstations (human–machine interfaces, HMI) is improving, while maintainability is the least addressed. A question exists if there are adequate cognitive standards related to controllability and system safety that help the users to understand new technology through digitalization, supporting adequate HMI.

TABLE 7.1

ABS Rules, Guides and Guidance Notes Related to Human Factors in Ship Design (American Bureau of Shipping, 2020)

Pub#	Title
86	Application of Ergonomics to Marine Systems
94	Bridge Design and Navigational Equipment/Systems
97	Risk Assessment Applications for the Marine and Offshore Oil and Gas Industries
102	Crew Habitability on Ships
103	Passenger Comfort on Ships
116	Review and Approval of Novel Concepts
117	Risk Evaluations for the Classification of Marine-Related Facilities
119	Ergonomic Design of Navigation Bridges
122	Alternative Design and Arrangements for Fire Safety
141	Fire-Fighting Systems
145	Vessel Maneuverability
147	Ship Vibration
151	Vessels Operating in Low Temperature Environments
154	Means of Access to Tanks and Holds for Inspection
163	Crew Habitability on Workboats
170	Rapid Response Damage Assessment
185	Integrated Software Quality Management (ISQM)
191	Dynamic Positioning Systems
201	Ergonomic Notations
209	Noise and Vibration Control for Inhabited Spaces
247	Habitability of Industrial Personnel on Accommodation Vessels
250	Application of Cybersecurity Principles to Marine and Offshore Operations – Cybersafety Vol 1
251	Cybersecurity Implementation for Marine and Offshore Operations – Cybersafety Vol 2
252	Data Integrity for Marine and Offshore Operations – Cybersafety Vol 3
278	Ergonomic Container Lashing
307	Guide for Smart Functions for Marine Vessels and Offshore Units

IMPLEMENTATION OF HUMAN FACTORS IN SHIP DESIGN

Investigations in offshore supply vessels (OSVs) operating in Norwegian Sea showed that lots of initiatives have been addressed in some of the new builds.

In a review, a number of field researches of human factors in ship design have been documented. The Royal Navy Habitability Survey presents valuable evidence of personal priorities and preferences for habitability features in warships based on a literature review, initial interviews and a pilot survey. Adequate levels of privacy and facilities for both individual and social relaxation were considered as important aspects in the ship's accommodation (Strong, 2000). A comparative study on the accommodations in royal naval and merchant naval fleets was conducted (Hardwick, 2000) by visiting ships and submarines and interviewing the crew. Suggested factors include the drive toward cabin-based accommodation for all crew, increased space for sleeping and personal stowage, improving working environment conditions

TABLE 7.2
IMO Publications Relevant to Human Factors in Ship Design

Assembly Resolutions (RES)	
A.342(IX)	Recommendation on Performance Standards for Automatic Pilots
A.468(XII)	Code on Noise Levels on Board Ships
A.601(15)	Provision and Display of Manoeuvring Information on Board Ships
A.708(17)	Navigation Bridge Visibility and Functions
A.817(19)	Performance Standards for Electronic Chart Display and Information Systems (ECDIS)
A.861(20)	Performance Standards for Shipborne Voyage Data Recorders (VDRs)
A.947(23)	Human Element Vision, Principles and Goals for the Organization
Maritime Safety Committee (MSC) Resolutions	
128(75)	Performance Standards for a Bridge Navigational Watch Alarm System (BNWAS)
137(76)	Standards for Ship Manoeuvrability
190(79)	Performance Standards for the Presentation of Navigation-Related Information on Shipborne Navigational Displays
IMO Circulars, MSC Circulars	
587	Life Saving Appliances
601	Fire Protection in Machinery Spaces
616	Evaluation of Free-Fall Lifeboat Launch Performance
645	Guidelines for Vessels with Dynamic Positioning Systems
834	Guidelines for Engine-Room Layout, Design and Arrangement
846	Guidelines on Human Element Considerations for the Design and Management of Emergency Escape Arrangements on Passenger Ships
849	Guidelines for the performance, location, use and care of emergency escape breathing devices (EEBD's)
982	Guidelines on Ergonomic Criteria for Bridge Equipment and Layout
1002	Guidelines on Alternative Design and Arrangements for Fire Safety

(noise and temperature) and provision of other facilities. An ethnographic approach (Lützhöft, 2005) pointed out illumination problems on the bridge, displays that were too bright and could not be dimmed, equipment that was not attached properly that the operator must put duct tape on and similar problems on 15 vessels that she visited.

Another investigation was pursued (Andersson & Lützhöft, 2007) using inter-views, field studies and questionnaires regarding environmental conditions in the engine department, ergonomic issues, engine and control room layout and technical interfaces. Deficiencies in the engine room on a merchant ship that did not comply with ergonomic principles as well as OHS requirements were reported.

Similar surveys were performed on seven Swedish merchant vessels (Grundevik et al., 2009) evaluating the design of engine control room (ECR), the layout, consoles and workstations. The result shows that the ECR design was not developed sufficiently to meet the demand and less in accordance with the technological progress. Ergonomic issues were reported, such as insufficient leg space, the position of the consoles and visibility problems. Defective hardware components, software bugs and defective software were among the most common system/equipment failures mentioned.

A 3D computer model was utilized to review a new US Navy ship (Dalpiaz, Emmrich, Miller, & McQuillan, 2005). They found that incorrect height/orientation for equipment, machinery and other manually operated equipment as the most common mistakes (20%). Stairs, ladders, steps and walkways designs were also found to be poorly designed (17%). Other deficiencies covered inaccessibility to valves, hand wheels and hand pumps (15%), incorrect control panel (i.e. controllability), console design, control and display designs (8%), and problems with access and personnel movement (6%).

Implementation of Human Factors in Ship Design –
Case Study: Offshore Supply Vessel

An OSV can be described as the workhorse of the oil and gas business in Norway. It is a combination of a bulk vessel, a general cargo vessel, a container vessel and a tanker with some extra capabilities like firefighting and oil recovery. The operational effectiveness, low fuel consumption, low emissions and safety are important factors in OSV design (Blenkey, 2004).

Field research of human factors was performed on OSV design in Norway (Rumawas & Asbjørnslett, 2011a; Rumawas & Asbjørnslett, 2013) using qualitative approach. Two advance vessels were studied: OSV A and OSV B. OSV A represents the regular OSV design with the bridge located in the front while OSV B offers an alternative design by putting the bridge at the back (Figure 7.1).

OSV was seen as an advanced design. Many improvements in ship design were initiated and tested in this OSV setting. The OSV design required a degree of sophistication unheard of previously in much of the marine world (Gibson, 2007).

Results from the observations indicated that human factors have been addressed in aspects of OSV design related to habitability, workability and controllability. Some improvements were made with respect to previous incidents, for example, the OSVs are designed with high bulwarks or side walls to avoid water on deck that previously moved containers causing injuries. Both vessels surveyed are built with excellent standards of accommodations. Each crew member sleeps in his/her own cabin equipped with television set, wireless internet connection, table, leather sofa, wardrobe, shower and toilet. Many new systems are digital, computerized and automated. While seafarers around the world felt frustrated for sailing a vessel which was

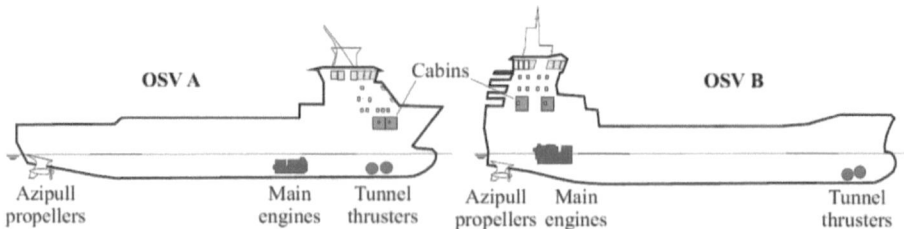

FIGURE 7.1   Two different offshore supply vessel (OSV) designs (Rumawas & Asbjørnslett, 2013). Reproduced with permission from the American Society of Naval Engineers.

designed with no crew input or whatsoever (Squire, 2007), most senior officers and engineers on OSV A and B were involved in the process of designing and building the vessel.

However, the crews know less of their vessel and what happened behind the screens. They can no longer rely on their senses to understand the state of their ships. The way they operate the vessel changes drastically. There is no wheel to steer the ship and no traditional engine telegraph to regulate the speed on these OSVs. The crews operate their vessel using joystick, mouse, track ball, buttons, keyboard and touch screen. Therefore, special adaptation and familiarization are required due to the novelty of the systems. Simulator-training has become necessary to operate these OSVs. Some ergonomics-related issues were found: one engine console was built without leg space, control buttons were located too far to reach from the crew's normal position so they must bend over or reach backwards to operate certain functions.

When the field surveys were conducted both vessels were in the process of maturing new technologies. On OSV A they were developing their DP system and on OSV B they were testing a hybrid power system. As new technologies were in their development stage, errors, miscalculations and flaws took place.

New kinds of problems were exposed, related to the digitalization and human factors such as incompatibility issue, operating system expired, overloaded system and hang, data invalidity, too much information presented on a screen and bugs. Flooding of alarms, abundant communication and procedures were also identified as issues on both OSVs.

Several cases can illustrate this. One officer told his story when he was operating OSV A on DP alongside an installation, and he sensed that the vessel unexpectedly began to move toward the installation. He shut down the DP system, took over the control manually and backed the ship away from the installation. One hose that was still connected snapped off. Later, the investigation revealed that the system assumed the vessel position was 100 m away from the installation while in reality it was 'only' 20 m. The manufacturer explained: "... this has happened only once on the entire DP X equipped fleet. We have found the root cause for this, and implemented a solution for it. This failure will not happen again." The statement implies that the particular condition was not identified when the system was finalized and it took place. An accident with a high hazard potential may have occurred. Fortunately, there was a human operator who acted as the barrier at that time. The officer's experience on this DP incident is an example of how handling of cues and sensemaking can play an important part to avoid major accidents at sea. Of course, not all incidents can be detected and followed-up by the operator (– as later described in *Sjoborg* case below).

Another example from OSV B illustrates the importance of resilience. The ship just finished loading and unloading process besides one installation and was preparing to maneuver to the next one. The officer on the bridge requested more power from the thrusters. Unfortunately, the system could not handle the power request and blackout occurred. The vessel was running on liquefied natural gas, which is less responsive to variations in the power requirements. When the bridge demanded power, the system automatically tried to switch over to diesel but failed. Fortunately, OSV B is equipped with a DYNPOS-AUTRO, i.e. class 3 DP. The vessel has sufficient redundancy, i.e. loss of position should not be caused by any single failure (including a completely burnt fire subdivision or flooded watertight compartment). Therefore, when blackout occurred, redundancy took over and the system went back to normal in a very short time (in seconds). The incident did not escalate. From the human operator's perspective, the crew knew that there was something wrong with the vessel, they were taken by surprise, but the system re-adjusted itself back to normal without the crew interference since it was safe by design. In normal ships without class 3 DP capabilities, it will be slower for the crew to bring back the system alive.

LESSONS LEARNED FROM AUTOMATION AND DP INCIDENTS

Over the period of 2001–2010, there were 26 collisions recorded between visiting vessels and facilities in the Norwegian shelf (Kvitrud, 2011; Oltedal, 2012; Petroleum Safety Authoritiy Norway, 2011; Vinnem, 2014). The following six of the collisions are considered high risk and safety critical. These cases demonstrate that there was some kind of vagueness regarding the state of the vessel (autopilot/DP), and then an emergency situation emerged, the operator tried to intervene, and in most cases their efforts to stop the ship actually made the outcome worse.

Three of the accidents involved autopilot and one involved DP system. Those accidents can be categorized as man–machine interface issues and thus related to human factors issues. The operators did not recognize that the system was active and could not take over control of the ship The system did not automatically disengage when overridden.

As we can see, the first accident recorded in the area was in 2004, and the standard was modified in July 2010 (DNV, 2010) as part of the revision in nautical safety (Section 6, Steering Control System). There was a time lag of 6 years on which similar mistakes occurred over and over and ended up in collision with the platform. Since 2010, the design of the console has been improved, the operator can take over control with one single action.

A DP-related accident took place just recently in 2019, involving a collision between the *Sjoborg* supply ship and *Statfjord A* during loading/discharging (Petroleum Safety Authority Norway, 2019). On 6 June 2019, 22:10, *Sjoborg* set in position on *Statfjord A* and performed discharging fresh water and diesel oil, and loading and unloading deck cargo. On 7 June, 01:04, alarms on integrated automation system (IAS) screen showed warnings: "FAULT IN B.O.S.S SYSTEM PS" and "FAULT IN B.O.S.S SYSTEM SB". These warnings were disregarded and not perceived as critical by the officer. From 01:14 to 01:49, a number of DP alarms appeared, were acknowledged but returned. Such alarms were considered "normal" by the crew during DP operation. 01:49, DP system changed to move vessel 6 m forward for access to deck cargo. IAS alarm showed: "BT1 AUTOSTOP", DP alarm: "TUNNEL BOW 1 NOT READY". IAS alarm showed: "BT3 AUTOSTOP", DP alarm: "AZIMUTH BOW 3 NOT READY". About 01:50, *Sjoborg* lost heading and position. Two of three bow thrusters dropped out. Vessel drifted toward *Statfjord A*. First officer attempted to switch *Sjoborg* to partly manual positioning. 01:51, *Sjoborg* hit *Statfjord A*.

This DP-related incident experienced by *Sjoborg* is rather different from the one on OSV A. The system on OSV A detected that the installation was further than its real location. The vessel then moved to adjust her location. The officer sensed that

TABLE 7.3

Safety Critical DP Collisions 2001–2010

Date	Collision between (Vessel) and (Installation)	Cause and Descriptions
7 March 2004	*Far Symphony* and West Venture	The vessel was heading toward the installation. The officers did not recognize that autopilot was engaged and could not take over control of the vessel in the safety zone. Instead, in their attempt to stop the vessel they increased the speed of *Far Symphony* hitting West Venture at 3.7 m/s.
2 June 2005	*Ocean Carrier* and Ekofisk 2/4 P bridge	The 1st officer navigated the vessel in poor visibility due to fog. The captain entered the bridge with the vessel passing the safety zone at 10 knots. Misunderstandings occurred as to who was responsible for navigation. The captain tried to slow down the vessel when he saw the platform but it was too late. *Ocean Carrier* hit the Ekofisk bridge at 3 m/s
13 Nov 2006	*Navion Hispania* (tanker) and Njord B	The tanker was in preparations to start offloading with the platform. Polluted fuel and clogged filters caused blackout, the tanker lost most power but one propeller. The DP system should be able to handle the situation but had an unrevealed fault. It was kept in "Autopos" mode. The crew tried to avoid collision but the situation escalated instead. The vessel hit the installation at a speed of 1.2 m/s.
18 July 2007	*Bourbon Surf* and Grane	The platform was identified as a target for the autopilot. The officers misjudged the ship's speed and distance to the platform; they left the bridge and did not keep a proper lookout. When they returned to the bridge it was too late to stop the vessel. They managed to reduce the speed, but *Bourbon Surf* hit the Grane at slightly less than 3.5 m/s.
6 June 2009	*Big Orange XVIII* and Ekofisk 2/4 W	The vessel was heading to the 2/4-X platform on the Ekofisk to perform well stimulation. The autopilot was engaged when the vessel entered the safety zone. The captain could not override the autopilot to control the vessel. The vessel managed to avoid Ekofisk 2/4-X, 2/4-C and Flotel COSLRigmar before she finally hit Ekofisk 2/4-W at a speed of 4.5–4.8 m/s.
18 Jan 2010	*Far Grimshader* and Songa Dee	The vessel was working on the lee side of the installation and was asked to move to the windward side due to the crane situation on the rig. When maneuvering, the vessel used extreme amount of power, causing the deck lights to go out, which the crew interpreted as power supply loss. The pitch control was then set to zero. The vessel was too close to the platform; one of the propellers was caught in a wire attached to the facility's anchoring. The vessel lost control and hit the installation repeatedly.

there was something unexpected with the behavior of the vessel and interfered. On *Sjoborg* a number of abnormalities have been accepted as "normal" and acknowledged by the crew. Further investigations revealed that the underlying causes were poor installation of equipment components causing network failure, in combination with alarms that were challenging to understand. Loss of network frequency measurement on the main switchboard activated the load-reduction mode, limiting the thruster output to only 15%. Nonconformity between DP commands and feedback from the thrusters took place, ended up in automatic shutdown of the thrusters, leading the vessel off course. Then the crew realized the problem and tried to interfere. But, it was too late. Thus, there is a question related to controllability and safety of the design of the control system. It is difficult for the crew to detect any peculiarities behind their screens, especially those related to digitalization. When warnings and alarms have become so common it becomes improbable for the crew to distinguish the safety-critical message. The crew can recognize that there is something wrong with their system when a more perceptible sign takes place, like power cut.

The alarms were probably not designed in accordance with accepted best practices alarm standards such as EEMUA 191 (EEMUA, 2013). The understanding of the alarms, alarm text description and design of the systems should be improved if human factors design standards had been followed.

QUANTITATIVE SURVEY OF HUMAN FACTORS IN SHIP DESIGN

Following up the field research that was conducted on the two OSVs in the Norwegian Sea described a quantitative survey was performed and reported (Rumawas & Asbjørnslett, 2015a, 2015b, 2016). Questionnaires and daily diaries were developed based on the existing human factors framework published by the LR (2009). Several factors like noise, temperature and motion were recorded. The survey revealed that the elements of "human factors" in ship design are quantifiable and measurable. It indicates that design can have a substantial influence on human factors assessment especially in habitability and workability. Good habitability of ship could reduce motion sickness incidence, fatigue and sleep disturbances on board. In turn, all these influence the operator's performance.

It is also revealed that some of the existing standards that govern noise (Det Norske Veritas, 2009b; International Maritime Organisation, 1981), motion and slamming (Graham, 1990; NATO, 2000; NORDFORSK, 1987) are too lax that neither they influence comfort nor safety. For example, the highest noise level measured in the cabin on OSV A was 56.7 dB (A) while on DP operation. Standard regulation in Norway defines that noise in work areas should be lower than 55 dB, and to avoid sleep disturbance, indoor guideline values for bedrooms are 30 dB L_{Aeq} for continuous noise and 45 dB L_{Amax} for single sound events – ref WHO (World Health Organization) (Berglund, Lindvall, & Schwela, 2009). The noise level conditions inside the cabin were uncomfortable due to a high level of screeching noise produced by the bow thrusters located not too far away. It was difficult for normal people to sleep in such a condition. The standards for seafarers allow noise level up to 60 dB (A) in cabins.

Another example, the operation criteria related to motion allows "heavy manual work" to be done when the vessel undergoes a vertical acceleration up to 0.15 g

(RMS) and 4.0° (RMS) amplitude of roll motion. Observation on board during heavy weather in the Norwegian Sea showed that it was impossible for anybody to stand still on a vessel in 14 m wave height. Physical measurement at that time indicated that the maximum vertical acceleration was 0.149 g (RMS) and maximum roll amplitude was 0.54° (RMS). It is obvious that these standards need to be revised to induce comfort as well as to ensure safety. In real life, the seafarers adjust the way they operate the vessel and their behavior on board. When the sea is high, the crew will try to find an alternative route that may be further but calmer or they can reduce the speed of the vessel and adjust the heading to reduce motion and slamming.

The quantitative survey confirms some aspects of human factors presented by the LR like controllability, workability and habitability. Other aspects seem to be weak. New dimensions appear: "cargo facilities", and "reliability automation and maintainability".

HUMAN FACTORS IN SHIP AS A SAFETY-CRITICAL SYSTEM

A study was conducted to develop a model to examine how ship accidents can be analyzed from the human factors perspective, given that a critical incident has already occurred (Rumawas & Asbjørnslett, 2010b, 2011b, 2014b). The focus of the study is on the operator's role. The hardware reliability perspective was adopted where the ship is considered as a safety-critical system to be protected by barriers. The crew is modeled as active barriers with different functions: perception, decision and action. A Markov model is utilized to describe different states of the crew on the ship. The highest state is 4: the crew performs the task correctly. The lower states are: the crew fails to monitor the situation (3), then fails to make the correct decision (2) and fails to perform the proper action (1). Two conditions are defined: normal condition (N) and extreme condition (E)[1]. Accidents usually happened in the latter, just as demonstrated in the section Lessons Learned above (Figure 7.2).

In reliability engineering, safety instrumented system (SIS) is defined as an independent protection layer that is installed to mitigate the risk associated with the operation of a hazardous system (Rausand & Høyland, 2004). Most of the time, the system is passive until a threatening situation takes place. The SIS model consists of a number of sensors, a logic solver and actuators (see Figure 7.3). The same logic is implemented toward ship operation by putting the crew as the SIS, meaning that they should be able to sense the hazard, to analyze the situation, to make a proper decision and to execute the right action. A mathematical model to estimate the probability of failure in an emergency situation is proposed (PFE). A parameter is defined for the survivability of a ship, given that a critical incident has taken place. Survivability, $S(t)$, is defined as the function of previous knowledge (S_0) combined with the accumulation of adaptation, on board learning processes (γ) and formal trainings or assessment (δ).

$$S(t) = S_0 \sum \gamma_i \tau + \sum \delta_i \qquad (7.1)$$

where τ is time between training or between assessments.

---

[1] Refer to the original paper for detailed description of the model.

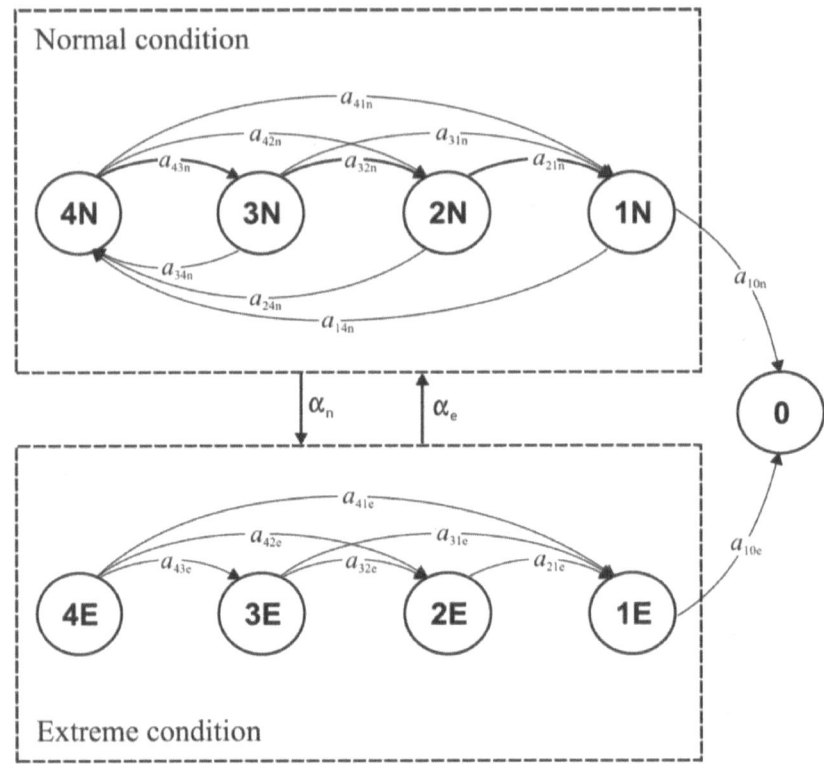

FIGURE 7.2   Markov diagram for ship operation (Rumawas & Asbjørnslett, 2011b).

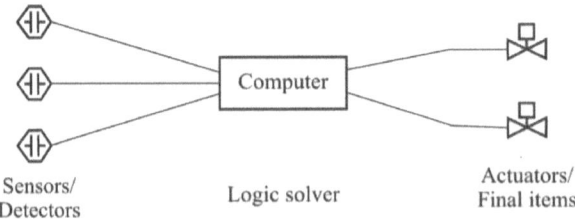

FIGURE 7.3   Safety instrumented system (SIS) model (Rausand & Høyland, 2004).

Thus, the probability of the crew failing to respond in case of emergency can be calculated:

$$PFE = 1 - S(t) \tag{7.2}$$

When a candidate is hired on a ship, he or she is not completely fit for the position. The crew has a certain level of knowledge and competence based his or her previous education, training and experience (denote as S_0). Nevertheless, every vessel is

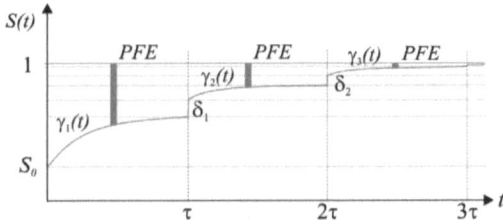

FIGURE 7.4 Survivability and probability of failure on emergency (PFE).

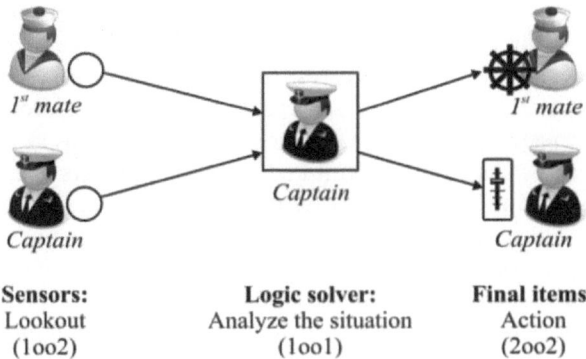

FIGURE 7.5 Crew modeled as safety instrumented system (SIS) in ship operation (Rumawas & Asbjørnslett, 2011b).

unique; therefore some orientation and adjustment will be required. The first time the crew is onboard, he or she will learn a lot (γ_1) for the period of τ (ref Figure 7.4). In addition, various training (δ_i) is expected to rise the capability of the crew to a certain level. The second time the crew is sent to sea, he or she will learn more (γ_2) but not as much as the first time. Figure 7.4 shows the concept of PFE. When the crew is new onboard, his or her knowledge of the vessel is rather limited. Therefore, the probability of the crew to fail in handling emergency on demand is relatively high. The longer the crew works and gets to know the vessel better, the lower the PFE would become.

An example of the crew modeled as barriers is illustrated in Figure 7.5, representing the bridge operation. Two officers are usually on watch on the bridge. They work as a parallel system when doing the lookout. It means that only one officer is required to detect any deviation on the vessel or in the surroundings (1oo2: one out of two). The captain in the middle acts as a logic server, analyzing the situation and making a proper decision. It is assumed that he performs the function by himself (1oo1: one out of one). In this model, his decision should be executed by both officers: the 1st mate and the captain must perform each duty properly (turn the wheel and adjust the lever) for the objective to be accomplished (2oo2: two out of two).

These methods were applied to evaluate ship–platform collisions cases (Rumawas & Asbjørnslett, 2011b, 2014b), and the results show strong benefits for diagnosing and evaluating accidents from a human factors perspective as well as for training purposes.

The study suggests that the crews' awareness of potential hazards and their knowledge of the ship (i.e. support of sensemaking and controllability) are very important factors that would determine the outcome.

CONCLUSION

A discussion of human factors consideration in ship design has been presented in this chapter. Research shows that human factors have increasingly been considered in the development of ship design, especially in OSV design in Norway. Several standards of how to address human factors in ship design is available in abundance – although some of the standards need to be revised. In maritime industry the implementation of human factors standards is not compulsory; it is quite different from the petroleum industry operating in the same area. In some cases, strong initiatives from the cargo owners (shippers) or the customers were witnessed as a driving factor to address human factors in the design. The pace of lessons to be learned in maritime industry is different from many other industries. As an example, it took more than 6 years and three accidents that were related to autopilot in the Norwegian Sea until the rules were revised.

In traditional ships where most of the equipment is mechanical and hydraulic, the crew usually understands what is going on the ship. In modern ships where most systems are digital and computerized, the crew is losing some of their understanding of what is going on behind the screens and systems. Coupled with imperfect automation and immature technologies, the crew is to some extent left in a difficult situation, especially when a hazard transpires in extreme or restricted condition. Thus, controllability and safety should be more in focus. There is a need to focus on how sensemaking is taking place and how systems are giving coordinated cues to the users. It is a systematic weakness that there is poor communication between the users and the designers and also between the manufacturers and suppliers, and that human factors still do not have a strong standing in the shipping industry.

It is recommended to consult and involve the users in the process of designing and improving a ship. Human factors expert involvement in the process is strongly supported. A comprehensive test and user acceptance of safety critical systems on board is recommended, besides the validations as required by the authorities.

REFERENCES

American Bureau of Shipping. (2020). *ABS Rules, Guides and Guidance Notes*. Retrieved from https://ww2.eagle.org/en/rules-and-resources/rules-and-guides.html

Andersson, M., & Lützhöft, M. (2007). Engine Control Rooms - Human Factors. *Paper presented at the International Conference on Human Factors in Ship Design, Safety and Operation*, London.

Ashe, G., & Lantz, J. (2003). Classification and Regulatory Requirements. In T. Lamb (Ed.), *Ship Design and Construction* (Vol. I). Jersey City, NJ: The Society of Naval Architects and Marine Engineers.

Baker, C. C., & Seah, A. K. (2004). Maritime Accidents and Human Performance: The Statistical Trail. *Paper presented at the MARTECH 2004*, Singapore.

Blenkey, N. (2004). OSV designers eye the future. *Marine Log, 109*(Compendex), 21–27.

Calhoun, S. R., & Stevens, S. C. (Eds.). (2003). *Human Factors in Ship Design* (Vol. I). Jersey City, NJ: The Society of Naval Architects and Marine Engineers.

Dalpiaz, T. M., Emmrich, M., Miller, G., & McQuillan, D. (2005). Conducting a Human Factors Engineering 3-D Computer Modeling Ship Design Review. *Paper presented at the RINA, Royal Institution of Naval Architects International Conference - Human Factors in Ship Design, Safety and Operation*, February 23, 2005–February 24, 2005, London.

Dekker, S. (2005). *Ten Questions about Human Error: A New View of Human Factors and System Safety*. Mahwah, NJ: Lawrence Erlbaum.

Det Norske Veritas. (2009a). *Class Notations Rules for Classification of Ships* (Vol. 2011). Høvik, Norway: DNV.

Det Norske Veritas. (2009b). *Rules for Classification of Ships Comfort Class: Special Service and Type Additional Class* (Vol. Part 5 Chapter 12). Høvik, Norway: Det Norske Veritas.

DNV. (2010). *Rules for Classification of Ships Newbuildings, Special Equipment and Systems Additional Class Nautical Safety* (Vol. DNV Pt 6 Ch 8). Høvik, Norway: Det Norske Veritas.

EEMUA. (2013). *Alarm Systems - A Guide to Design, Management and Procurement*. EEMUA Publication (Third ed., Vol. 191). https://www.eemua.org/Products/Publications/Digital/-EEMUA-Publication-191.aspx

Gale, P. A. (2003). The ship design process. In T. Lamb (Ed.), *Ship Design and Construction* (Vol. I). Jersey City, NJ: The Society of Naval Architects and Marine Engineers.

Gibson, V. (2007). *The History of the Supply Ship*. Aberdeen: Madrila.

Graham, R. (1990). Motion-Induced Interruptions as Ship Operability Criteria. *Naval Engineers Journal, 102*(2), 65–71. doi:10.1111/j.1559-3584.1990.tb02556.x

Grundevik, P., Lundh, M., & Wagner, E. (2009). Engine Control Room - Human Factors. *Paper presented at the International Conference of Human Factors in Ship Design, Safety and Operation*, London.

Hardwick, C. (2000). A Comparative Assessment of Priorities for Accommodation Standards between Royal Naval and Merchant Naval Fleets. *Paper presented at the International Conference on Human Factors in Ship Design and Operation*, London.

International Association of Classification Societies (IACS). (2004). *What Are Classification Societies*. www.iacs.org.uk

International Maritime Organisation. (1981). *Code on Noise Levels on Board Ships* (Vol. Res A.468(XII)). London: IMO.

International Maritime Organization. (2004). *Human Element Vision, Principles and Goals for the Organization* (Vol. A.947(23)). London: IMO.

International Maritime Organization. (2012). *Casualty Statistics and Investigations Loss of Life from 2006 to Date*. London: IMO.

International Maritime Organization. (2020). *Introduction to IMO*. Retrieved from http://www.imo.org/About/Pages/Default.aspx

Kvitrud, A. (2011). Collisions between Platforms and Ships in Norway in the Period 2001–2010. *Paper presented at the International Conference on Ocean, Offshore and Arctic Engineering, OMAE 2011*, Rotterdam, The Netherlands. http://kvitrud.no/OMAE2011-49897%20 COLLISIONS%20BETWEEN%20PLATFORMS%20AND%20SHIPS%20IN%20 NORWAY.PDF

Lloyd's Register. (2008). *The Human Element: An Introduction*. London: Lloyd's Register.

Lloyd's Register. (2009). *The Human Element Best Practice for Ship Operators*. London: Lloyd's Register.

Lützhöft, M. (2005). Human Integration of Bridge Technology. *Paper presented at the International Conference on Human Factors in Ship Design, Safety & Operation*, London.

Lützhöft, M., & Dekker, S. W. A. (2002). On Your Watch: Automation on the Bridge. *Journal of Navigation, 55*(GEOBASE), 83–96.

McCafferty, D. B., & Baker, C. C. (2006). Trending the Cause of Marine Incidents. *Paper presented at the International Conference on Learning from Marine Incidents*, London. http://www.eagle.org/eagleExternalPortalWEB/ShowProperty/BEA%20Repository/References/Technical%20Papers/2006/TrendingCausesMarineIncidents

Miller, G. E. (1999). Human Factors Engineering (HFE): What It Is and How It Can Be Used to Reduce Human Errors in the Offshore Industry. *Paper presented at the Offshore Technology Conference*, Houston, Texas.

NATO. (2000). *Standardization Agreement (STANAG): Subject: Common Procedures for Seakeeping in the Ship Design Process*. Brussels, Belgium: NATO, Military Agency for Standardization.

NORDFORSK. (1987). *Assessment of Ship Performance in a Seaway*. Copenhagen: Nordforsk.

North West European Area. (2009). *NWEA Guidelines for the Safe Management of Offshore Supply and Rig Move Operations*. London: Chamber of Shipping, Danish Shipowners Association, Netherlands Oil and Gas Exploration and Production Association, Norwegian Ship Owners Association, United Kingdom Offshore Operators Association.

Oltedal, H. A. (2012). Ship-Platform Collisions in the North Sea. *Paper presented at the European Safety and Reliability Conference*, Helsinki, Finland. https://core.ac.uk/download/pdf/30807266.pdf

Petroleum Safety Authority Norway. (2011). *Risk of Collisions with Visiting Vessels*. Retrieved from http://www.ptil.no/news/risk-of-collisions-with-visiting-vessels-article7524-79.html

Petroleum Safety Authority Norway. (2019). *Investigation of Collision between Sjoborg Supply Ship and Statfjord A on 7 June 2019*. Retrieved from https://www.ptil.no/contentassets/b545c860789f4a61bb6f684cbdfc295a/2019_764_eng-rev-rapport-equinor-granskingsrapport-statfjord-a-sjoborg-kollisjon.pdf

Rausand, M., & Høyland, A. (2004). *System Reliability Theory: Models, Statistical Methods, and Applications*. New York: John Wiley & Sons.

Reason, J. (2000). Human Error: Models and Management. *Western Journal of Medicine, 172*(6), 393–396. doi:10.1136/ewjm.172.6.393

Ross, J. M. (2009). *Human Factors for Naval Marine Vehicle Design and Operation*. Surrey: Ashgate.

Rumawas, V., & Asbjørnslett, B. E. (2010a). A Content Analysis of Human Factors in the Design of Marine Systems. *Paper presented at the International Conference on Ship and Offshore Technology (ICSOT)*, Surabaya.

Rumawas, V., & Asbjørnslett, B. E. (2010b). A proposed model to account human factors in safety-critical systems. In B. M. Ale, Ioannis A. Papazoglou, & Enrico Zio (Eds.). *Reliability, Risk and Safety Back to the Future* (pp. 1760–1766). Boca Raton, FL: CRC Press.

Rumawas, V., & Asbjørnslett, B. E. (2011a). Offshore Supply Vessels Design and Operation: A Human Factors Exploration. *Paper presented at the European Safety and Reliability Conference (ESREL)*, Troyes.

Rumawas, V., & Asbjørnslett, B. E. (2011b). Survivability of Ships at Sea: A Human Factors Perspective. *Paper presented at the ERGOSHIP 2011*, Gothenburg.

Rumawas, V., & Asbjørnslett, B. E. (2013). Exploratory Surveys of Human Factors on Offshore Supply Vessels in the Norwegian Sea. *Naval Engineers Journal, 125*(2), 69–85.

Rumawas, V., & Asbjørnslett, B. E. (2014a). A Content Analysis of Human Factors in Ship Design. *Trans RINA International Journal of Maritime Engineering, 156*(Part A3), 251–264. doi:10.3940/rina.ijme.2014.a3.299

Rumawas, V., & Asbjørnslett, B. E. (2014b). Survivability of Ships at Sea: A Proposed Model to Account for Human Factors in a Safety Critical System. Trans RINA *International Journal of Maritime Engineering, 156*(2), 137–148. doi:http://dx.doi.org/10.3940/rina.ijme.2014.a2.284

Rumawas, V., & Asbjørnslett, B. E. (2015a). Human Factors Evaluation in Ship Design: A Case Study on Offshore Supply Vessels in the Norwegian Sea, Part I: Theoretical Background and Technical Constructs. *Naval Engineers Journal*, in press.

Rumawas, V., & Asbjørnslett, B. E. (2015b). Human Factors Evaluation in Ship Design: A Case Study on Offshore Supply Vessels in the Norwegian Sea, Part II: Multivariate Analyses and Structural Modelling. *Naval Engineers Journal*, in press.

Rumawas, V., & Asbjørnslett, B. E. (2016). Human Factors on Offshore Supply Vessels in the Norwegian Sea - An Explanatory Survey. *International Journal of Maritime Engineering, 158*(A1), 1–14. doi:http://dx.doi.org/10.3940/rina.ijme.2016.a1.329

Squire, D. (2007). 'Fit for Purpose' - Keeping the Crew in Mind. *Paper presented at the International Conference on Human Factors in Ship Design, Safety and Operation*, London.

Strong, R. (2000). RN Habitability Survey: Ship Design Implications: Some Important Social and Architectural Issues in the Design of Accommodation Spaces. *Paper presented at the International Conference on Human Factors in Ship Design and Operation*, London.

Vinnem, J. E. (2014). *Offshore Risk Assessment: Principles, Modelling, and Applications of QRA Studies* (Vol. 3rd Ed). London: Springer.

Vossen, C., Kleppe, R., & Hjørungnes, S. R. (2013). Ship Design and System Integration. *Paper presented at the DMK Conference.* https://www.researchgate.net/publication/273026917_Ship_Design_and_System_Integration

8 Sensemaking in Practical Design
A Navigation App for Fast Leisure Boats

T. Porathe
Norwegian University of Science and Technology

CONTENTS

INTRODUCTION: THE CHALLENGES OF HIGH-SPEED NAVIGATION

A Long Time Ago

When I was a boy, I used to spend my summers at my grandparents in a small coastal village in the west of Sweden. My grandfather was a fisherman and much of my time there was spent with him at sea. Just outside the harbour pier there was a reef, normally hidden by only a few inches of water. Some 20 m further out was the lateral buoy warning for the danger. However, all native fishermen took a shortcut inside the buoy, aware of the exact location of the shoal. Only strangers followed the rules of the road and took the buoy on the right side. One day, when my grandfather was in his 70s, he hit that reef. It was no big deal; the boat heaved over in the water and was then washed across the rock by the wake. But it was a big embarrassment. For 50 years he had gone in and out through the pierheads almost every day without problem, and then a few moments of inattentiveness in an area he knew so well. Does this tell us something about human behaviour?

Three Years Ago

A warm summer night with heavy rain in the archipelago of southern Norway. After midnight, a water scooter at full speed is heading home after a late-night concert in a small coastal town. But the driver never returns home to the summer cabin where the family wait. The next morning the scooter is found crashed on a small island some distance from the home. The driver is dead, instantly killed on impact with the rocky island.

The police states that alcohol, darkness and bad visibility have played a role in the crash: "It is very hard to manoeuvre at sea in darkness. It can be different from time to time even if you go the same route", said the search and rescue leader at the local police district (Verdens Gang, 2018). One of the challenges of high-speed navigation is short decision time. We will never know the full reason why this accident happened.

Two Years Ago

Just before 2 o'clock in a dark and moonless night in August 2019, a fast leisure boat crashed into a small island in a fjord in middle Norway. The boat was home-bound through a fjord and hit a small island just outside the fairway in high speed, 36 knots. Both the driver and the passenger were badly injured by the impact and later died. The driver was well known in the area and had done the trip many times.

FIGURE 8.1 To the left, the crashed water scooter on the accident scene in 2018. To the right, the crashed open speed boat on the island where it ended up in 2019. (Images courtesy of VG, 2018 and NRK, 2020.)

The accident commission noted that there had been Snapchat activity on the driver's mobile phone in the seconds leading up to the crash (Statens Haverikommisjon for Transport, 2020). A few seconds of inattentiveness could very well be the crucial factor leading to the accident (Figure 8.1).

BACKGROUND

FINDING YOUR WAY AT SEA

Navigation is a Greek word stemming from *navis* (a ship) and *agere* (to drive). Navigation is about knowing where you are and knowing what way to take to reach your goal. In olden and even modern days, *pilots* are used to navigate the ship. Maritime pilots are people with local knowledge about underwater dangers and how to get from one place to another. Geographical data were collected and recorded first in itineraries and sailing directions and then as nautical charts, paper and nowadays electronic. Today also mariners unfamiliar with an area can find their way. Finding your own position by referencing landmarks on islands and coastlines has today been replaced by a position plotted by global navigation satellite systems. However, navigation in unfamiliar waters is difficult even with the help of nautical charts and automatic position fixing. Even if you know the way you need to go in the chart you still have to deduce steering marks in the terrain leading to your goal. To do that you need to pay attention to visual as well as other cues, focus on the task, use implicit, explicit and prospective memory resources. In short you need to pay attention and make sense of many cues in order to perform safe navigation. It costs cognitive resources. This is something humans can do very well – but also very badly as inattentiveness is part of the human condition.

HUMAN FACTORS

The ability to focus and sustain attention on a task is crucial for the achievement of one's goals. Although attention span is a complex concept and measures depend on a lot of different things, a common agreement among researchers is that the time span healthy teenagers and adults can concentrate to handle tasks without being distracted is limited to 10–20 minutes (Wilson & Korn 2017). Navigating in your own backyard is a piece of cake and very easy. This is what my grandfather felt in the story above. This might be what the driver of that water scooter and that fast leisure boat also felt navigating in well-known areas. Accident investigations often talk about complacency, a feeling of calm satisfaction with your own abilities or situation especially when accompanied by unawareness of actual dangers or deficiencies. Fifty years of successfully sailing in and out of the port had made my grandfather complacent to the danger posed by the shoal outside the pier.

The accidents described in the beginning of this paper could be examples of "human error". According to Donald Norman (2013), one category of such errors is *slips*. Slips occur when a user is on mental "autopilot" and takes wrong actions pursuing a goal, typically when the user does not fully devote his or her attention to the task at hand. The question a designer asks himself here is if there is any simple help that can be provided to avoid these kinds of accidents in the future? The necessary data are already available: the position of the boat and future position within a reasonable time frame and chart data showing water with enough depth to sail in. This information could be available on a chart application ready to be used by a user. And there is in fact an abundance of such apps. Figure 8.2, left, shows a typical chart plotter used by leisure mariners (the same was used in the accident boat in the last story above). The problem is only that you still need to pay attention to the information shown on them, and in a context as shown in Figure 8.2, right, that attention needs to be spent on driving. A typical chart plotter is simply not useful in many small and fast leisure boats.

SENSEMAKING

"Sensemaking can be seen as the process to establish situational awareness based on cues" (Kilskar et al., 2020). This is your conscious and focused navigator comparing the planed route on the map with landmarks in the archipelago around us. But when you are sitting on a water scooter in 40+ knots with both hands clung to the

FIGURE 8.2 To the left, a chart plotter used in many leisure crafts including the one in the last accident narrated above. To the right a water scooter. Typically, with speeds between 40 and 60 knots. (Images courtesy Garmin and Sea-Doo.)

handles not to fall off, or in moments of inattention in a fast-moving leisure craft it is a different thing. In this situated context, sensemaking is "the process of searching for a representation and encoding data in that representation to answer task-specific questions" (Russel et al., 1993). Russel et al. continues "Different operations during sensemaking require different cognitive and external resources. Representations are chosen and changed to reduce the cost of operations in an information processing task". So, while drivers in slow weather-sheltered cabin cruisers might benefit from the protection and time to study information on a chart machine, the speed, vibrations and time constraints of very fast boats and water scooters necessitates other solutions. This was the one of the motivators of the *Sikker kurs* project. The other was the problem of inattentiveness.

Sikker Kurs

In 2016, the Sikker kurs project started, financed by the Norwegian Coastal Administration and Geomatics Norway AS. The purpose was to develop an application for ordinary smartphones to increase safety and possibly decrease the number of groundings by leisure boats in Norwegian waters. The project was a proof-of-concept demonstrator, coordinated by Geomatics Norway. The design was made by the author, working at the Norwegian University of Science and Technology (NTNU) in Trondheim, and technical implementation was conducted by Combitech AB in Linkoping, Sweden. Other partners in the project were the Norwegian Hydrographic Office (Kartverket) and the Norwegian Maritime Authority (Sjofartsdirektoratet). It was decided that the project would use the human-centred design process (HCD) in ISO 9241-210 (ISO, 2015) and International Maritime Organization's (IMO) guideline on HCD (IMO, 2015).

Concept

Our goal was to make a navigation aid that would support sensemaking during sea trips and reduce the risk of groundings (increase safety) in small boats. We would do this by designing a smartphone app that would warn the driver of an imminent grounding danger 30 seconds before impact. Because of the challenging environment on many fast leisure boats (e.g. a water scooter), the warning should be aural (and could potentially trigger an automatic engine cut). After stopping the craft, the driver should have an opportunity to see and understand why the warning had been given and also find a way out of the situation (Porathe & Ekskog, 2018). In order to test this concept, we started to develop a proof-of-concept.

METHOD

Human-Centred Design (HCD)

The point of HCD is to ensure user-driven development and good usability by including the end users early in the process and keep them involved during the whole design. This is done in an iterative process with four steps according to ISO 9241-210:

1. Understand the context of use by field studies and interviews with the users.
2. Specify the user and organisational requirements.
3. Produce a design solution, this will be the prototype.
4. Evaluate the design against requirements. Here, the prototype is tested on the end users.

The findings are then brought into a new iteration of the design process resulting in a new, improved prototype. The process is then iterated until the application meets the requirements.

Test Area and User Group

Before a user group could be recruited, a location had to be decided. One way would be to look for an area with a large amount of leisure boat traffic. However, the availability of very detailed bathymetry was necessary and a difficult problem. The Norwegian Hydrographic Office offered an area in Søre Sunnmøre, a district south of Ålesund on the Norwegian west coast which had been declassified and could be used. A central municipality in this area was Ulsteinvik, which was to become the centre of the project. We needed to find local leisure boat mariners. A letter was sent out to 30 pleasure craft clubs in the district informing about the project and asking about participation in development and testing of the application. Unfortunately, only six leisure boaters responded, all male, all relatively experienced and in the age group of 60+. But luckily for us, these end users have helped us a lot with testing during the years.

Understand the Context of Use and User Requirements

A first focus group meeting was held in Ulsteinvik in January 2017. The users were interviewed about their experience with leisure craft navigation and the proposed concept of using a smartphone as a means of preventing groundings was discussed. The group concluded that the idea was interesting and that there was a need for a safety device alarming if the boat was approaching unsafe depths. The group agreed on a prioritised list with different possible features (Porathe & Ekskog, 2018).

Alarm

The phone should sound an alarm a configurable time before the boat went aground. The application should be automatically started in the background when a boater steps onto his boat, so that he or she does not forget to start the application. The time should be short so that the number of false alarms in narrow archipelagos would not be annoying and thus making boaters turn off the alarm (which is often the case with the look-ahead-sector in professional shipping). The default setting was agreed as 30 seconds, and the procedure of the boater should be to immediately stop the boat on alarm. The alarm should be silenced by picking up the phone and clicking on the warning icon shown. The alarm should also be silenced by slowing down to a configurable maximum speed (default three knots) to allow boats to make landfalls or approach a jetty without getting an alarm.

NoGo Areas

When the phone is picked up and the alarm silenced, the screen should show "NoGo Areas" in red overlayered on the camera image (so called augmented reality, AR). These NoGo areas are the polygon inside a configurable depth contour. The default was the 3-m contour, but ideally, any depth should be able to be picked based on the current draft of the boat, plus a safety margin. Ideally, the depth alarm should also compensate for the current tidal situation based on tide tables or real-time tide gauges. The user should be able to see these NoGo areas all around by pointing the smartphone camera.

Landmark Names

Conspicuous landmarks around the boat should be named by overlaying text on the camera image. Examples of such conspicuous landmarks could be names of islands, shoals, buoys, beacons and mountaintops (the area is very mountainous). Much time on board a small craft is spent trying to find buoys and beacons. An overlaid pointer should show their position to aid visual search. To avoid cluttering, the names and pointers could be toggled on and off by tilting the camera (slightly up turns text on and vice versa).

Air Draught

An alarm similar to the grounding alarm could be configured for sailing boats with a mast height that is higher than the span of oncoming bridges and power lines.

Fairways and Planned Routes

Official fairways should be shown as an AR "carpet" rolled out on the water in the camera image. Also, individual routes planned in a chart program and imported into the phone could be shown in the same manner. This feature must be able to be turned on and off to avoid cluttering. This requirement was later dropped for the tested prototype due to time constraints.

TECHNICAL PROTOTYPE DEVELOPMENT

After the meeting with the user group, discussions started about the technical implementation and what could be achievable within the time and budget available. Of the five prioritised solutions suggested by the user group, the first four were selected for development.

The Android Platform

We decided to make the test implementation on the Android platform because Combitech had earlier experience in this platform, had available equipment and the relative ease with which test implementations could be distributed without being passed by the AppStore (for Apple's iOS), thus giving us a quicker development cycle. Recently, the app has also been developed for iOS.

NoGo Areas and Alarm Execution

Part of an Electronic Navigational Chart (ENC) was imported into a database in the phone's memory. From the ENC, only the polygons making up the area with a water depth of less than 3 m at chart datum were kept. These polygons made up the "NoGo Area" that was used to alarm the navigator for grounding. Ideally, we would have NoGo polygons for every decimetre, which would be turned on and off depending on the set draught of the boat and the tidal situation. However, this would require large memory storage or a constant online connection, so we decided to have just one NoGo depth of 3 m for the test. The Norwegian Hydrographic Office delivered the necessary depth contour with a high-resolution horizontal grid of 1 m. The internal map would consist of polygons marking water depths between 3 m and 0 (the beach line).

The timed alarm function was implemented using a vector extending from the present position in the direction of the current course. The length of the vector was dependent on the speed and the alarm time set. In the default setting, the alarm was set to be triggered 30 seconds before the boat "grounded" (passed into the 3 m NoGo area polygon). At 10 knots, the length of the vector would be (10 knots * (1,852 m/3,600 seconds) * 30 seconds) =154 m. The length and direction of the course-speed vector was calculated from recent satellite positions. The precision was dependent on the position rate the phone could muster, which in general was one position per second (1 Hz). The alarm would be triggered when a course-speed vector intersected with a NoGo area polygon.

The air draught alarm was treated the same way using the same course-speed vector intersecting a safety rectangle extending 15 m on both sides of bridges and power lines. The set mast height would then be compared against the maximum air draught allowed as stated as an attribute to the safety rectangle. In the test area, there was only one power line and no bridges.

The Augmented Reality (AR) Layer

The NoGo area polygon map was to be shown on top of camera image at the correct position. The polygons should apparently be "floating" on the surface of the water. In order to do this, the map had to be georeferenced and projected using a virtual camera positioned in virtual space as the real camera was in the real space. This projection is a standard virtual reality (VR) operation conducted in real time taking the virtual camera's height over the water (preset to 2 m), direction (from the phone's compass) and field of view (preset to match the device's camera) as in-parameters.

The course-speed vector was also made visible and projected into the camera view: white when not in alarm mode but changing colour to red when an intersection had taken place and the alarm was triggered. It was then red as long as it was intersecting with the NoGo polygons, thus visualising the alarm state, also when the aural alarm was silenced. The initial intersection point was shown by an arrow.

The stability and precision of the satellite positions and the compass heading from the internal phone sensors was an area of concern. The course-speed vector triggering the alarm was created by extrapolating present course and speed into the future. Low-pass filters were applied to these values to avoid large jumps due to unstable satellite fixes. This was done to reduce the risk of false collision alarms. The point of view in the polygon map was also dependent on the satellite-based present position, but the direction of the camera (which was independent from the course-speed vector of the boat) was relying on a compass direction from the phone's internal magnetic compass. We had little experience of the precision of these two sensors, which might also be dependent on local conditions in the area for the test. However, to anticipate possible problems with the compass, we made it possible to shut down this sensor and use the course-speed vector as direction for the virtual camera in the augmented reality layer, then assuming that the camera was fixed in a forward-looking manner (for example, on the windscreen).

The only text-based information we considered we had time and resources to implement was the pointer for navigational marks. The position of all buoys and marks in the test area was collected in a list. We did not succeed in populating the list with all the marker names in time, so the markers in the tests prototype mostly showed "POI" for point of interest.

RESULTS

The first iteration of the prototype was tested during a technical test in Ulsteinvik with two people from the user group on 8 May 2017. The full user test was conducted a month later with the six people from the user group (Porathe & Ekskog, 2018).

Technical Test

For the technical test in May, a relatively complex 5.8 nautical miles long track was drawn in an ENC (see Figure 8.3). This track could be negotiated in a little more than an hour at a moderate speed of 5 knots (not to take any risks should the prototype prove unreliable).

For the test, we used a 7-m leisure boat owned by a member of the user group. He also had very good local knowledge, which would be a safety barrier against

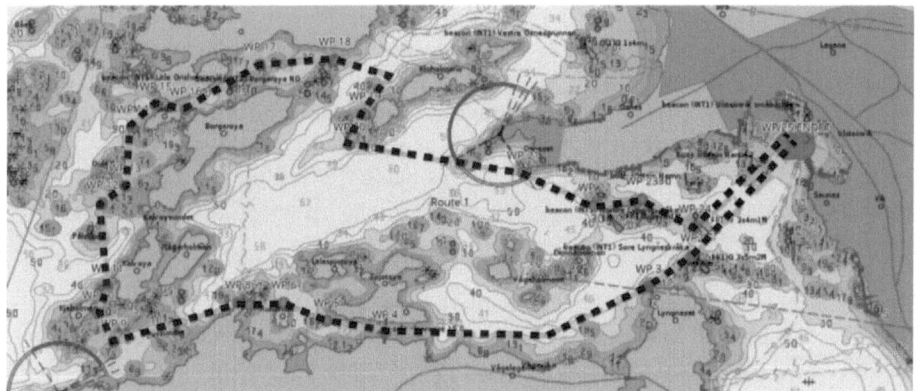

FIGURE 8.3 The test track outside Ulsteinvik in western Norway. (Map courtesy Kartverket.)

FIGURE 8.4   The test application on the smartphone (Lenovo Phab 2 Pro) during the pilot test. To the right, the boat's reference stationary chart plotter. (Photo courtesy of the author.)

unintentional grounding should the prototype fail. The boat was also equipped with a stationary chart plotter which was used as a reference system (see Figure 8.4).

The prototype software was tested on two phones: a Samsung Galaxy S7 and a Lenovo Phab 2 Pro. We found no differences in behaviour between the two phones. Some problems with the fluctuating AR layer are described below.

User Test

The final user test was held in Ulsteinvik on 14 June 2017. The same test track as in May was used and all six of the original users were present on the 15 m M/S Legona used during these tests. The boat made the passage at about 5 knots speed in just over an hour and the prototype was tested on the two phone types mentioned above. Below are the results of this user test (Porathe & Ekskog, 2018).

Alarm Execution

The function to automatically turn on the application when leaving port was not developed for this first prototype. The application was manually turned on when the test run commenced. When the course-speed vector intersected the NoGo area, the alarm was triggered, both while the phone was "sleeping" in the pocket or (as in Figures 8.5–8.7) when the phone was used to actively monitor the water ahead of the boat. By touching the stop sign, the alarm is acknowledged and silenced, and the stop sign disappeared. However, the vector remained red as long as it was intersecting a NoGo area. This feature worked perfectly as designed and the comments from all the users were very positive.

NoGo Areas

The AR layer was projected over the camera image based on a virtual camera positioned by latitude and longitude from the phone's GNSS sensor, and the virtual camera's direction was based on input from the phone's internal magnetic compass. Both these sensors had fluctuations as opposed to the camera image, which of course moved only when the phone moved. This resulted in smaller or larger fluctuations of the AR layer over the camera image. The AR layer with the NoGo areas, course-speed vector

FIGURE 8.5 Screen dump from the Galaxy test phone. The projected NoGo areas in red. The 30 second course-speed vector in white just before the alarm is triggered. The pointers showing three points of interest (two of which is hidden behind the island). (Photo courtesy of the author.)

FIGURE 8.6 Screen dump from the Galaxy test phone. The grounding alarm has been triggered with both an aural and a visible alarm. (Photo courtesy of the author.)

FIGURE 8.7 A very narrow passage on the test track. The distance between the 1-m shoal and the small skerry is 33 m (in the chart, left). Right, the app view entering the narrows (northbound). (Photo courtesy of the author.)

FIGURE 8.8   The picture shows the offset of the augmented reality (AR) layer with the red NoGo areas. There is a vertical offset and a horizontal and fluctuating offset due to noise in the phone's internal compass. However, users judged it acceptable during the tests. (Photo courtesy of the author.)

and POI pointers would float or jump in the image, mostly in the horizontal plane. These fluctuations would be more or less prominent depending on factors such as if the camera was being panned and/or magnetic disturbances in the boat or in the area. The sensitivity to magnetic disturbances is illustrated by this example: one phone tested had a leather cover that could be closed over the screen with a magnetic lock. This lock jammed the compass causing the AR layer to become unreliable.

The horizontally fluctuating AR layer was the one disappointment in an otherwise successful test. Ideally, the layer with its added information should be steady and "glued" to the camera image, in the test prototype it jumped or sailed some 5°–10° to either side of its intended position. However, the user group judged it to be within reasonable limits. This was because the inside of the NoGo areas was visually easy to pair together with the island's beach line, making the fluctuations "some kind of visual expression of uncertainty" (user comment). In Figure 8.8, an offset to the right and slightly up can be seen. The beach line of the island and the front beach line in the inner hole of the red NoGo areas match. Note also that the NoGo areas behind the island are visible which they should not be. Theoretically, they could be clipped using an invisible 3D terrain model in some future version of the app. This 3D terrain model could then be shown during darkness and fog when the camera showed nothing. However, the most important thing was that the triggering of the grounding alarm function was not affected by the fluctuations due to the magnetic compass. The alarm computation was done entirely in the map layer using the relatively more stable GNSS position.

Points of Interest

The pointers to named points of interests (for example, lighthouses, buoys and other marks) are potentially beneficial as a second source of information to cross check the visual integrity of the system. However, this feature was not tested as we did not get access to names of the markers in the area (which were not present on the chart). In the prototype, most marks only carried an anonymous "POI" label. This feature will be investigated further.

Survey

After the test voyage and a short debriefing, the six users answered some questions in a small survey. The first question was whether they thought that the tested prototype could have any favourable effect on boat navigation. On a scale from 0 to 100, where 0 was "no favourable effect" and 100 "large favourable effect", they were asked to indicate their answer with a cross. The mean result of all six users was 83, close to "large favourable effect".

The second question dealt with the usability of the prototype application. On the same type of scale from 0 to 100, where 0 was "simple to use" and 100 was "difficult to use", they were asked to mark their answer with a cross. The mean result from the six users was 13, clearly on the "simple to use side".

They were also asked to comment on the prototype and asked if they missed any functions. Three answered "no", one gave no answer and the remaining two made these comments: "The matching between the AR layer and the camera image could be better", "Automatic Identification System (AIS) data could be added", "Some adjustments and it will be fine", "Get it out as soon as you can, new versions can come later".

During a concurrent television interview (NRK, 2017), one of the users commented on the alarm function: "I am often out sailing in my boat and when tacking we often want to use the water between the islands as much as possible, and then often go close to land. If we could get an alarm by a buzzer in the pocket instead of having to constantly look on our navigator screen, that would be great".

Ongoing and Future Development

The proof-of-concept was successful, but after 2017 the development stopped lacking funding. However, the user group in Ulsteinvik continued testing the app, now named *GrunnVarsel*. The area was very limited to the archipelago west of Ulsteinvik but the user group managed to uncover some important problems not found during the initial user test. Figure 8.9 shows a screen from the test videos made by the user group. One such important problem was that when the side of an island fell steeply into the sea, there was no NoGo area polygon generated and thus no warning. In

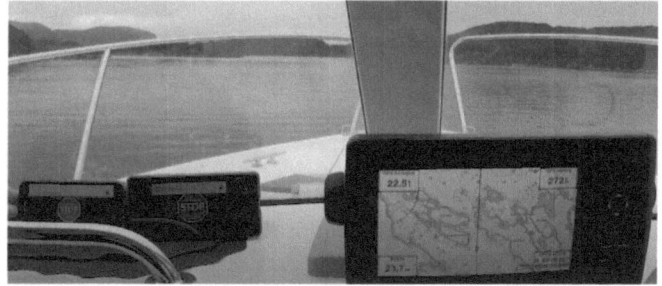

FIGURE 8.9 Tests in October 2019. To the left two tested smart phones where the grounding alarm has just been triggered. To the right the chart plotter shows speed position, heading and distance to the triggering depth curve. (Photo courtesy of Harald Notøy and Leidulf Garshol.)

these cases, there needed to be a safety margin manually added to the beach line (and also around buoys and markers moored on water deeper than 3 m).

The app has now also been ported to the iOS (Apple) platform. In 2020, the decision was taken to start a second phase of the development with the same actors and financed by the Norwegian Coastal Administration. This time, the test area will be outside Tønsberg on the Norwegian south coast and the test will focus on technical benchmarking and reliability of the app.

DISCUSSION

The intention of this project has not been to develop an application to replace traditional navigation methods but to create a "last line of defence" against accidents. However, it will be difficult to prevent a few boaters from using it as a sole means of navigation. The question is: If we develop a "simple, stupid" application, which facilitates boating for leisure mariners without navigational training, – do we then lure new "unfit" groups of people out on the sea, which in the end might lead to more accidents? And, do we contribute to the de-skilling of leisure mariners?

Let us make a parallel with professional navigation. Traditionally, ship's positions were acquired by measuring the angles to the sun or terrestrial landmarks. After some calculations, you obtained a "historical" position, where the ship recently was. This position was then manually plotted onto the paper chart. There were abundant opportunities of making errors during the measurement, the calculations or during the plotting, let alone that overcast days or bad visibility sometimes made measuring the sun height impossible.

When the radio-based Decca and Loran systems and later the global positioning system came, the measuring process was automated and only the manual plotting into the chart remained until Electronic Chart Display and Information System (ECDIS) allowed the officer to have the ship's position automatically plotted on the chart in real time. In 1989, the IMO issued the first provisional performance standards for ECDIS (IMO, 1989) and in 1995 the US Coast Guard presented an early human factors study (Smith et al., 1995). It concluded that "ECDIS had the potential to improve upon the safety of navigation, compared to conventional procedures", and that "there was strong evidence that the use of ECDIS increased the accuracy of navigation, […], and reduced the proportion of time spent on navigation, with a corresponding increase in the proportion of time spent on the higher risk collision avoidance task. In addition, ECDIS was shown to improve geographic 'situational awareness' and to reduce navigation 'errors' " (Smith et al., 1995, P.VIII). Spontaneous comments such as "Navigation goes away as a task" were made by the participants.

However, this was achieved at the cost of what we call de-skilling. No longer did the mariners need to train their skills in taking sun heights with the sextant or bearings with a pelorus. They became more dependent on the automatic systems. In an article in the *Journal of Navigation*, Edmund Hadnett (2008) from the Port of London Authority reacted to the de-skilling of navigators in dependence on modern bridge technology leading to "over-confidence in situation awareness, encouraging individuals to take far greater risks than was previously the case where a good look-out and a safe speed were intrinsic parts of watch-keeping". Hadnett (2008) concluded that "The drive to improve safety at sea by the introduction of electronic navigational equipment to enhance situation awareness and assist the watchkeeper has unwittingly compromised safety standards by reducing the core competences that were demanded of previous generations and engendering the undesirable human trait to select the easiest option".

Furthermore, de-skilling continues, now the ECDIS itself has become too complicated. In the foreword of the UK Maritime Accident Investigation Board's (MAIB) report after the Ovit grounding in the English Channel 2013, the UK Chief Inspector of Marine Accidents wrote "This is the third grounding investigated by the MAIB where watchkeepers' failure to use an Electronic Chart Display and Information System (ECDIS) properly has been identified as one of the causal factors." (MAIB, 2014. P.1)

However, although the observations that the de-skilling amongst professional navigators are undoubtedly true, the safety and reliability of modern shipping keep improving from year to year. To provide a perspective, it is interesting to note that in the 3 years 1833–1835, on average 563 ships per year were reported wrecked or lost in the United Kingdom alone (Crosbie, 2006). The world fleet of tankers, bulk carriers, containerships and multipurpose ships, which have risen from about 83,000 ships in 2011 to more than 98,000 in 2020 (UNCTAD, 2020). The global number of reported total shipping losses of over 100GT declined during 2019 to 41 – the lowest total this century and a close to 70% fall over 10 years (Allianz, 2020). So, although automation has led to de-skilling, it has also led to safer shipping. The question now is, can the same argument be made for technology in leisure navigation? I would say yes and argue that a simple, automated tool, warning leisure mariners against grounding, will potentially result in fewer accidents if properly developed in the process of going from prototype to product.

CONCLUSION

Sensemaking in small and fast leisure crafts works differently than in the protected environment of slower and lager boats. In this study, a simple, smartphone-based safety application was developed and tested. Leisure boaters often have a limited knowledge of navigation according to accident statistics, and the application was designed to be easy to use and understand without prior knowledge. It worked in two ways: (1) In a "turned off" mode in the pocket, the phone would give an alarm 30 seconds before the boat entered into "dangerous waters" (depth less than 3 m). The boat owner was then expected to immediately stop the boat. (2) Picking up the phone, the owner could look through the application's camera view and see red "NoGo Area" polygons overlaid on the camera image. By looking around, he or she could then detect navigable water and continue the voyage.

The application contained a high-resolution map of the 3-m depth contour extracted from a nautical chart. This map was then projected on the camera image's "egocentric view" of the surroundings, thus bypassing the potentially cumbersome mental rotations a human navigator has to do when comparing a traditional exocentric map with the world around. This would facilitate use by inexperienced boaters.

The application was tested on a small group of six Norwegian, all male, all experienced, leisure craft mariners. The size and configuration of the test group limits the generalisability of the results, but the group had highly positive views of the tested prototype, which encourages continued work on this project.

Future work includes adding some limited features asked for by the user group while still maintaining a simple and easy-to-use app. The most prominent new feature will be the ability to import a pre-planned route from a nautical chart application (or an official route from the Coastal Administration) and show this route in the AR layer overlaid in the camera image, thus not only showing dangers to navigation but also offering way-showing.

The intention of this experiment was user experience (UX) and to find out if such an egocentric AR application would be beneficial and would potentially be used by leisure mariners in an archipelago setting. Precise technical benchmarking and testing of different smartphone brands potentials and problems were not undertaken, but is the task for an ongoing project.

The intention is not to replace the normal navigation procedure, but to add an extra safety layer.

The initial goal with this design project was to see if we could manage to develop a safety tool that would allow fast leisure boaters to benefit from the digitalisation of navigation that has been going on for many decades. This digitalisation has resulted in a dramatic decrease in accidents with commercial ships. The exposed environment in many small and fast boats has prevented sensitive and voluminous equipment to be installed and read during voyage. And if such a tool would be found beneficial by the user group, the proof-of-concept with a very limited user group was quite successful and the project has commenced in a second iteration in the Tønsberg area in southern Norway in 2020.

REFERENCES

Allianz. (2020). Safety and Shipping Review 2020. https://www.agcs.allianz.com/content/dam/ONEMARKETING/agcs/agcs/reports/AGCS-Safety-Shipping-Review-2020.pdf [Acc. 2021-02-19]

Crosbie, J.W. (2006). Lookout Versus Lights: Some Sidelights on the Dark History of Navigation Lights. *The Journal of Navigation,* 59, 1–7.

Hadnett, E. (2008). A bridge too far? *The Journal of Navigation,* 61, 283–289.

IMO (International Maritime Organization). (1989). *Provisional Performance Standards for Electronic Chart Display and Information Systems (ECDIS).* NAV 35/WP.31989, IMO, London.

IMO (International Maritime Organization). (2015). *Guideline on Software Quality Assurance and Human-Centred Design for e-Navigation.* MSC.1/Circ.1512, IMO, London.

ISO (International Organization for Standardization). (2015). *ISO 9241-210:2010 Ergonomics of human-system interaction — Part 210: Human-centred design for interactive systems.* https://www.iso.org/standard/52075.html [Acc. 2021–02–19]

Kilskar, S. S., Danielsen, B. E., & Johnsen, S. O. (2020). Sensemaking in Critical Situations and in Relation to Resilience—A Review. *ASCE-ASME Journal of Risk and Uncertainty in Engineering Systems, Part B: Mechanical Engineering,* 6(1).

Marine Accident Investigation Branch (MAIB). (2014). *Report on the Investigation of the Grounding of Ovit in the Dover Strait on 18 September 2013.* MAIB, London.

Norman, D. (2013). *The Design of Everyday Things: Revised and Expanded Edition.* Basic Books.

NRK (2017). Distriktsnyheter Møre og Romsdal, reporter Arne Flatin.
NRK. (2020). Høy fart og alkohol kan ha medvirket til båtulykken hvor Malvik-ordfører døde. https://www.nrk.no/trondelag/hoy-fart-og-alkohol-kan-ha-medvirket-til-batulykken-hvor-malvik-ordforer-dode-1.15064816#:~:text=Tips%20oss!-,H%C3%B8y%20fart%20og%20alkohol%20kan%20ha%20medvirket%20til%20b%C3%A5tulykken%20hvor,august%20i%202019. [2021-02-19]
Porathe, T., & Ekskog, J. (2018). Egocentric Leisure Boat Navigation in a Smartphone-based Augmented Reality Application. *The Journal of Navigation*, 71(6), 1299–1311.
Russel, M.D., Stefik, M.J., Pirolli, P., & Card, S.K. (1993). *The Cost Structure of Sensemaking*. https://dl.acm.org/doi/pdf/10.1145/169059.169209 [Acc. 2021–02–19].
Smith, M. W., Akerstrom-Hoffman, R. A., Pizzariello, C. M., Siegel, S. I., Schreiber, T. E., & Gonin, I. M. (1995). *Human Factors Evaluation of Electronic Chart Display and Information Systems (ECDIS)*. United States Cost Guard Research and Development Center, New London, CT.
Statens Haverikommisjon for Transport. (2020). *Rapport Sjø 2020/04: Rapport om sjøulykke med fritidsbåt, Lokkarskjøret, Namsos, 1 August 2019*. Statens Haverikommisjon for Transport, Oslo.
UNCTAD. (2020). *2020 e-Handbook of Statistics*. https://stats.unctad.org/handbook/MaritimeTransport/MerchantFleet.html#:~:text=World%20fleet%20development%20and%20composition,recorded%20an%20especially%20rapid%20increase. [Acc. 2021-02-19]
Verdens Gang. (2018). Politiet mener Skofterud manøvrerte feil i mørket. https://www.vg.no/nyheter/innenriks/i/On7gmw/politiet-mener-skofterud-manoevrerte-feil-i-moerket [Acc. 2021–02–19].
Wilson, K. & Korn, J.H. 2007. Attention During Lectures: Beyond Ten Minutes. *Teaching of Psychology*, 34 (2): 85–89.

9 Unified Bridge – Design Concepts and Results

F. B. Bjørneseth
Norwegian University of Technology and
Science, Kongsberg Maritime

CONTENTS

BACKGROUND AND DRIVING FORCES

The increasingly advanced automation systems controlling modern sea vessels have led to more complex user interfaces. A typical operator must interact with several different systems, often with different interface styles, during an operation. Complex and multiple interfaces can cause cognitive overload if the operator is presented with excess information. The operator can also be physically affected if the equipment is poorly placed. Depending on the ship owner, the shipyard and the suppliers of equipment, the composition of the equipment in the operator station can vary considerably and is often ergonomically sub-optimal. From the user's part, the process of sensemaking becomes more important in such a complex environment. By sensemaking, we mean the dynamic, iterative process of observing, orienting and acting in a social setting, thereby creating a shared understanding (Kilskar et al. 2019; Weick, 1988).

This section examines the effect of a how a user-centred design process together with design thinking was a differentiator when designing a new ship bridge environment.

The concept development started by involving the user from the ideation phase and throughout the product development process towards a finished product. The final product was released into the market as a (unified) ship bridge environment designed with a holistic perspective including a redesign and rearrangement of the physical consoles, input devices and software interfaces located in the environment to support four design criteria: *safety*, *simplicity*, *performance* and *proximity*.

THE DESIGN CRITERIA AND SENSEMAKING

The ship bridge environment is a safety-critical environment related to operations/ navigation where errors may cause significant damage to vessel, crew and environment. The first of the main design criteria, *safety*, originates from the numbers presented that 75%–96% of maritime accidents can involve some sort of human error (Allianz Global Corporate & Specialty, 2012). The contemporary view is that human error is a consequence of deeper issues with the system. This can be a combination of issues such as poor design, poor training, mental overload, fatigue (Dekker, 2004). The aim was to address especially poor design and mental overload, hence increasing safety by providing the operator with a ship bridge work environment where during standard operations, the cognitive load on the operator was as low as possible. The operator could then spend time and effort on the ongoing operation, rather than to operate the vessel. Supporting this would leave the operator with a clear mind and a ship bridge environment that supports fast decision-making rather than providing the operator with increased workload as the environment gives room for interpretations that introduce misunderstandings and doubts during safety-critical events. The operator could then have increased *performance* during safety-critical situations. To fulfil this criterion, the ship bridge work surfaces needed simplification and decluttering, hence *simplicity*. When simplifying the environment in immediate vicinity to the operator, important functionality such as touch screens and operator devices could then be brought to a closer *proximity* of the operator. The design criteria support the human factor and sensemaking. Literature and previous research within other domains state that human factors and ergonomics (HF/E) research demonstrate that extreme levels of cognitive workload decrease an individual's ability to react to incoming information and increase the likelihood of human error (Nocera et al., 2007). In addition, an example from the power industry by Holzinger et al. (2012) illustrates how it is possible to reduce the complexity of user interfaces for safety-critical power-plant control systems. The operators need to make sense of the information presented to them; they need to perceive and interpret to be able to make decisions (Weick, 1993). The design criteria support this.

To fulfil the criteria, an iterative approach including insight studies, operator interviews and eye-tracking was selected. The continuous feedback loop from the crew on board the vessel using the ship bridge in daily operation was important to gain further insight and continue improving the concept. After 5 years in the market, a benchmark insight study was carried out (Danielsen et al., 2019), where the feedback from the vessel operators was positive, with reluctance to return to a more conventional ship bridge environment.

Human–machine interface (HMI) work has a long history in maritime settings but is often given low priority due to economic pressures and perceived increased development time. The economic aspects play an important role in a vessel's lifecycle and issues concerning HMI and usability are, in many cases, not a part of the discussion until late in the cycle when it is too late and too expensive to make vital changes to implement an optimal solution. An overall increased mental load when operating a system is tiring and leaves less mental capacity to handle safety-critical events. Poorly fitted equipment combined with low usability causes a long-term problem for the operators. The overall aim of this particular project was to lower the operator's cognitive load and make the workflow on the ship bridge more efficient.

The project's objective and main requirement were:

To increase operational safety of demanding offshore operations through:
  – A complete re-design of the ship bridge environment (levers, chairs, consoles and software interfaces), incorporating human factors, ergonomics and usability as the base foundation for development.
  – Introducing a more comfortable and safe working environment for the operators.

IN PERSPECTIVE

Traditional ship bridges are often cluttered with equipment, buttons and levers. The placement often depends on who arrived first to install their equipment at the shipyard. With no holistic focus on where to place equipment, it is either just randomly placed somewhere in the consoles, placed according to a setup from the ship design supplier without any thought of operational environment or placed according to. the wish of the captain on duty that day. When asked, the crew often reply that they have concerns, but that the concerns are just "silly details" or "luxury problems". It is worth reflecting on the fact that the operators in this study tended to downgrade their own perceptions, which can affect the safety on board and the ability to understand when "the point of no return" occurs when entering a critical situation.

However, it soon becomes clear that while the individual problems might be small, there are usually many of them, and as they start to pile up they add unnecessarily to the operator's mental workload. The sum of small insignificant issues makes holes in the barriers defending against accidents, thus the saying "For want of a nail a kingdom was lost" should be used here as well, all insignificant issues may become significant during an accident when time is limited and sensemaking must be performed based on many different cues. When taking all the silly details and luxury problems into account, it was possible to create a ship bridge concept that will improve sensemaking, operational safety and comfort on board during demanding offshore operations.

THE USER-CENTRED DESIGN PROCESS AND DESIGN THINKING

The methodology of design thinking and the process of user-centred design are two similar ways of emphasizing and understanding the user from the user's perspective. The user-centred design process is based on the ISO 9241-210:2019 standard that outlines the core principles of user-centred design but with a focus towards digital interfaces, human computer interaction and digital processes (ISO 9241-210, 2019).

When redesigning a ship bridge, the human computer interaction is an important part of the holistic user experience. To be fully able to understand and support the selected methodology, there are two additional standards that are important to lean on: ISO 6385 (ISO, 2004), which specifies the ergonomic principles intended as a guide for the design of work systems and as an extension, and ISO 11064-1–7, which has special focus on the case of designing control centres, which a ship bridge can be defined as. The combination of the two techniques, design thinking and the process of user-centred design, is therefore beneficial in this case: as will be explained below as a part of the design thinking process, a mixture of investigative and generative methods to get an understanding of what the user needs. To gather requirements at an early stage in an innovation project is challenging, as the requirements are unclear and the user does not really know what is important. The user-centred design process holds five different steps that includes:

1. Research phase –this stage typically consists of observation studies and interviews.
2. Concept phase – try and fail with different concepts, including a pre-phase of low fidelity prototyping using paper, cardboard and post-its.
3. Design phase – the low fidelity prototyping advances and develops into more mature prototypes as in this case utilizing polystyrene, 3D-printed prototypes, milled large scale plastic models a complete ship bridge setup ready to be tested in a simulated environment.
4. Develop phase – at this stage the product is ready to be materialized after reducing the risk after iterating through the three previous stages. In this case, building the first concept delivery at a workshop for the users to come and test it before being installing on board the real vessel was a valued process.
5. Test phase – the two first deliveries of the Unified Bridge were still called prototypes, however already tested through several iterations. The last phase was to test the concept in its real environment and evaluate it iteratively and continuously to improve and gather feedback for later versions of both software and hardware.

The user-centred design process will then iterate and continue until you have a product that fulfils the initial design goals and general requirements as outlined in ISO 11064-1–7.

When comparing the user-centred design process and design thinking, it seems similar and intertwined throughout the steps. Both are emphasizing with the end users and both are putting focus on the end user. However, when looking closely, one can distinguish that design thinking is a broader term and be applied to all products and processes. The user-centred design process focuses more on the digital perspective. Design thinking has focus on innovation, ideation and finding user-focused solutions as a basis for building and developing products and solutions. The user-centred design process is more focused on the actual creation of user-focused digital interfaces (ISO 9241-210, 2019; Browne et al., 2008).

For the Unified Bridge development, both were important to incorporate as the project and concept development had an important digital perspective with the redesigning of all digital surfaces and software applications to create an unified look and feel across all products delivered by this supplier as well as a high degree innovation and walking through unploughed fields that required an open mindset.

The development of the Unified Bridge is described through the five steps of design thinking:

1. Emphasize
2. Define
3. Ideate
4. Prototype
5. Testing

METHODS

EMPHASIZE

When redesigning a ship bridge environment from a human perspective, using design thinking and user-centred design processes, the most important element is to emphasize with the user. Emphasize by gaining understanding, obtaining knowledge and insight within the field of seamanship, by observing the ship bridge crew and doing interviews. The observations were carried out at sea during actual operations followed by semi-structured interviews with the crew (Bjørneseth et al., 2012). Obtaining knowledge about the daily routines on board is just as important as the observations and interviews. Complementing the observations and interviews by gaining an understanding of the shift rotations and the general life on board was a vital part of getting a holistic understanding and for an inexperienced external to be able to understand and emphasize with the crew on board.

Observations were also carried out in a ship bridge simulator again followed by interviews. The simulator observations had a more specific goal where looking into the actual operation was of interest. In the simulated operations, only one operator was present (Bjørneseth et al., 2014), and by using simulators, it was now also possible to include novice operators. Comparing the behavioural patterns of experts versus novices gave valuable insight into equipment placement according to the operator's eye movements and scan patterns. Eye-tracking equipment was utilized, which will be outlined in more detail in the sections below.

To add more strength to the comments from on-board ship bridge crew and onshore operators (simulator observations), unstructured interviews were also carried out at two additional vessels. The vessels were at the time of the interviews docked quayside, preparing for a new trip offshore, loading equipment and liquids for the nearby oilfields. The ship bridge crew was on duty at the ship bridge but was not occupied with safety-critical or demanding tasks. The crew was more relaxed on the ship bridge, monitoring the loading of the vessel. By observing and interviewing the ship bridge crew in a more informal setting, valuable insight where gathered concerning the crew's decision making process and influence on where to place different types of equipment. This social process of enquiry enables us to understand the needs of the crew from an user-centred design perspective – understanding the tasks of the sailors – and from a sensemaking perspective, i.e. how cues are shared and the social collaboration helps to create a common mental model of what is going on.

DEFINE

Asking the correct questions when assembling the information gathered in the previous phase, emphasizing, is important to get the correct angle to the "problem" and to define the problem and investigate how it might be possible to solve. During the define phase, the Unified Bridge design criteria evolved, which was to design for:

-Safety	- Simplicity
- Performance	- Proximity

Safety, where the aim was how it might be possible to develop a safer ship bridge environment than we have today.

Simplicity, where the aim was to look at how it might be possible to simplify the often over engineered solutions.

Performance, where the aim was to look at how it might be possible to make the ship bridge operator increase his/her performance by decluttering workspaces/ surfaces (both physical and graphical user interfaces) and alert management.

Proximity, where the aim was to look at how it might be possible to bring the most important functionality and operational surfaces closer to the user.

The next step to be able to define and build more detailed insight was to carry out task analysis on a selected set of tasks defined as important by the interviewed operators. The focus was mainly on the operational tasks that was defined as the safety-critical tasks. This included navigational tasks, transferring command from the forward bridge workstation to the aft bridge workstation during dynamic positioning operations when entering the 500-m safety zone surrounding offshore oilrigs, monitoring automation system and dynamic positioning system during cargo loading and offloading and alert handling. According to accident reports (Dhillon, 2007), the operator's cognitive load during standard and critical operations are of interest, as stress and cognitive overload is in many cases identified as one of the causes when accidents and incidents occur. It was therefore found beneficial to combine a lighter version of the hierarchical task analysis methodology, by also looking into the cognitive aspect of it. According to Howell and Cooke (1989) when advances in technology have increased and not decreased the mental demands on the user, including a cognitive task analysis at some of the tasks presented would be beneficial. In the maritime domain, as a world-leading equipment supplier, this was the key initiator for commencing on this project. The rapid technological development of equipment with short deadlines to deliver left equipment suppliers with little thought to the operators' well-being during the development phase (i.e. the classical technology driven approach and not the user driven approach). In many cases, the operator's opinion was only asked after the product had been finalized, purchased and installed on a vessel's ship bridge. This leaves the operator with little to no influence on the products that were to be important in the operators' success or failure when handling large vessels in harsh weather conditions during complex offshore operations.

The standard equipment supply chain and decision-making lies with the ship owner, the shipyard and the equipment supplier(s). This topic is a different discussion outside the scope of this chapter, however not without significance for today's lack of HF focus in the maritime industry. Typically, the ship owner identifies a need for a new vessel and asks for an offer from multiple shipyards. The shipyards assemblies offer from multiple suppliers where the supplier who can meet the requirement at the lowest cost can sell their product to the builder. The yard with the lowest cost wins the contract and can commence to build the vessel. There is no focus on a holistic perspective of what types of equipment should be fitted next to each other or where the equipment should be placed. The class requirements, which will be discussed later, are a guideline that leaves plenty of room for individual interpretations both from shipyard suppliers and surveyors when approving the vessel's certificates.

Looking into the task analysis (Kirwan and Ainsworth, 1992) when analysing the give and take command (command transfer), the five steps outlined by Marine (2014) were utilized during observations:

1. **The trigger points**. What triggers the operator to start their task?

   To use as an example, transferring command from forward ship bridge workstation to aft bridge ship bridge workstation. This action is triggered when the vessel reaches the 500 m safety zone around the oil rig and has been given permission by the rig to approach. However, the decision is not exclusively made exactly on this border. There is individual decision-making done by the officer on watch when this transfer is to be activated. What is known is that it does happen when the vessel is moving towards the oilrig within the safety zone. Weather conditions such as wind, waves and currents may affect the decisions made, where the decision requires skills and cognitive processing.

2. **Task goal and desired outcome.** When do the operators know when this task is complete? The command transfer starts when the operator (often the commanding officer) on the forward ship bridge workstation asks the operator on the aft ship bridge workstation if he/she is ready to take command of the vessel. When this is confirmed, the operator on the forward ship bridge workstation presses the command transfer button and the operator on the aft bridge workstation must accept the transfer. The system activates and signalizes both in its graphical user interface by illuminating the buttons on the levers indication that the aft workstation is in command. This is the cue for a completed task; however, testing that the vessel is in the new command position is common, by initiating action to the system to feel the anticipated vibrations and movements that belong together with operating a vessel.

3. **Base knowledge.** What do the operator need to know and are expected to know when initiating this task? The operator must hold knowledge concerning when it is appropriate to transfer the command according to surrounding circumstances such as weather conditions, navigational position and by communication with the oil rig. In an integrated operation, such as in this example, where the operator's origin is from different environments and cultures, the vessel can be considered as one culture and the oilrig another.

The different cultures need to collaborate towards a common goal, which can be challenging when they interpret their environments differently.

4. **Required Knowledge.** What do the operators need to know in order to complete the task? The operators need to know about the command transfer procedures, make sure it is according to the checklist, press the command transfer button at the workstation and acknowledge the transfer at the other workstation. This includes few and simple steps, however with great responsibility and demanding decision making for it to be a safe decision.

5. **Artefacts.** The tools and information the operators utilize during the task is the physical buttons and levers placed in the vessel's workstations, the communication equipment for communication with the oilrig and the DP checklist procedures. The communication with the oilrig and the equipment utilized to carry out this task can also be looked at as a separate operation that can be broken down into lower level tasks. To dive into that level of detail was, however, not important for this research project.

Investigating the different functions of each safety-critical piece of equipment was also included in the analysis. The example above described a command transfer utilizing the command transfer button and belonging software. To map the frequency of use of the particular function is also of interest. Three questions were asked:

1. How often do they use this function? Continue using the command transfer example, meaning how often do they transfer command from one ship bridge workstation to another?
2. Where is the equipment placed?
3. Is the placement sensible and clear to the operator in context of the above questions?

By looking at the interaction patterns, frequency of use and functional importance, it was possible to form an opinion on where the equipment should be placed according to the observations (Bjørneseth et al., 2012) and measurements done in the simulated study (Bjørneseth et al., 2014).

THE LUXURY PROBLEMS AND THE COGNITIVE LOAD

During the interviews and observations, none of the operators complained even though it was clear to the observers that workarounds and personal adaptations were common in the ship bridge workstations. Some examples were the use of blue magnets (from the white board), Figure 9.1, to mark the active pump in the emergency stop button panel. In addition, post-its and small tablecloths covered displays and monitors that were too bright during night sailing and also, ergonomic issues having to turn their bodies into unnatural positions to reach the internal telephone from the workstation chair. This was resolved in the finished concept, by providing the operator with a tidier console, where the internal telephone could be moved to an appropriate position within arm's reach. Before cleaning up the clutter, there was no space for the internal telephone, and it was placed at the far end of the console behind the operator.

FIGURE 9.1   Whiteboard magnet to visualize the active emergency stop button.

The typical question operators were asked:

What would you like to change in your ship bridge work environment?

The answer was in general:

If we want any changes? Well… I'm mostly happy. Yes, really! However, I'm not sure it is worth to mention… It's probably just a silly detail…maybe even a luxury problem?!

The operators adapted their behaviour and routines to the solutions presented to them on board, as they had no other option. Looking back in literature and using a practical example from daily life, it is possible to imagine that to search and find something important to you (when you are in a hurry), such as your keys, when the keys are hidden under some old magazines on the kitchen counter cluttered with toys, glasses, pots and pans, is a difficult task. Visual clutter causes an increased load on the working memory, the cognitive load, reducing the amount of cognition you have left to handle other tasks. In a safety-critical maritime environment, a high cognitive load during standard routine tasks leaves less cognitive load for the safety-critical moments, where decisions have to be taken quickly without having to search for a button, lever or having to read the labels or search through numerous software menus to understand which button to press or action to take. When piling up all the little details, all the little "luxury problems", it will in the end present a possible threat to the safety on board as these "little luxury problems" come in addition to more serious and well-known issues on board, such as alert handling and alert avalanches. The alert handling on board is often poor, as there can occur alert avalanches, where one root cause alert triggers and avalanche of other alerts. An example of this is when the vessel loses GPS signal. All equipment that relies on the GPS signal will then have their alert triggers to a cacophony of audible alerts and an alert list that are significant in length where the root cause is well hidden. The operators' level of experience will

FIGURE 9.2   Illustration of pre-redesign (to the left) and post redesign (to the right) of aft bridge workstation.

be important in how the situation is handled. An experienced operator will instantly know that there most likely is no danger and that the circumstances implies that the vessel is now in GPS shadow. An unexperienced/novice operator might experience increased stress levels and have less capacity to handle the vessel during a demanding operation (Figure 9.2).

This led the project into the next phase of the design thinking process, ideation.

IDEATION

Industrial designers who had been a part of the two first phases of the process, could also in this phase help with identifying problems in the current phase and sketching new solutions to the problems identified in the previous phases. The new solutions presented had to be grounded in one important requirement: "The concepts presented must be based on technology available today in the suppliers' market, not in the future". With this requirement, the new ship bridge was not a vision that possibly could be developed some time in the future but be ready for the maritime market within a time period of maximum 5 years. Hence, the concept could *not* lean on futuristic and unavailable technology to be successful.

During the ideation phase, three concepts were presented, and four main areas of change were identified that needed to be redesigned:

   1. Ship bridge consoles – the actual steel consoles where the equipment is to be mounted.

2. The operator chair – the chair is present on all bridge stations (forward, aft and wing stations) and is fitted with all relevant equipment mounted in the chair's armrests.
3. Control levers and joysticks – lever and joysticks that are to control the vessel's different systems such as propulsion (main propulsion/azimuths/tunnel thrusters, etc.), dynamic positioning joystick, winches and rudders.
4. Software interfaces – the software interfaces that momentarily have different interaction patters, navigation patterns, icons, etc. depending on which system the operator interacted with.

PROTOTYPE

In this phase, the design thinking process now merges even closer with the user-centred design process. The users had in the first phases been observed and interviewed to collect as much insight as possible. During the prototyping phase, the users should now be involved from the very beginning of the design process: starting with low-fidelity prototyping using cardboard, paper and polystyrene to investigate heights, view angles and equipment placement (Figure 9.3).

Several iterations of prototyping were carried out, advancing and changing the layout for every iteration in collaboration with the industrial designers and their

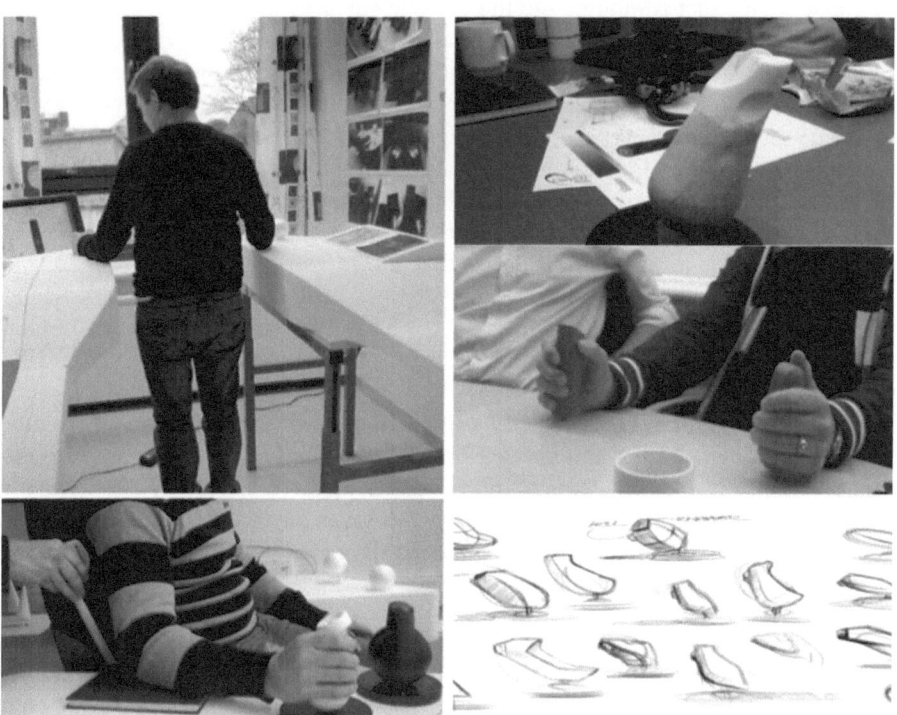

FIGURE 9.3 Prototyping activities.

concept drawings. The prototyping continued in four different projects, where one project concerned the actual consoles, their layout, size and ergonomics; the second concerned the levers, joysticks and monitors that were to be fitted in the consoles; the third project concerned developing the new ship bridge operator chair and the fourth project concerned redesigning the software interfaces. This project did however commence at a later stage after the physical prototypes were in place. The three projects concerning physical equipment had their separate development runs where experienced users gave feedback on designs drawings and prototypes continuously. In addition, the designs were tested with inexperienced users with no maritime background, mainly to gather general more informal feedback from a different point of view.

THE REQUIREMENTS

The user-centred design process emphasizes the need for collecting requirements. In innovation projects, the requirements are often not clear. The overall requirement was the project objective, as described above, to develop a safer and more efficient work environment.

When redesigning the input devices (levers, joysticks, etc.), gathering technical requirements from the product owners gave directions and limitations to the design.

An example of a main requirement was that all input devices bases (the part of the input device that is mounted to the console surface) should have standard measurements and shapes; in this case, a circular shape was selected as it saved console footprint. The footprint of the lever equals the space it requires on the surface of the console. In addition to saving footprint, the circular shape gave room for the necessary buttons each type of input device required. If an input device did not need all the buttons available, the spare buttons were blended but available for later extensions. The depth of the levers, meaning how deep into the console they reached after being fitted into the consoles, gave directions to the console design. One of the console requirements that emerged was that a slim and tidy visual expression was desired. To fulfil this requirement, the mechanics of the input devices also needed a redesign. Other important requirements were button shape, number of buttons, input device illumination and the icon design for the buttons had to be standardized across all devices.

The different requirement owned by the different projects influenced each other and created a demand for finding well-processed compromises. In some cases, the slim aesthetically pleasing design lines created in the consoles by the industrial designer have to give way for a more chunky design as the levers had to get fitted into the console at the correct placement, at just the right location for the operator to have his/her ergonomic requirements fulfilled.

EVOLUTION

The prototypes evolved in several iterations from the very low fidelity, low cost, low threshold type of prototypes to more high fidelity fully functional prototypes, and the console prototypes developed from paper, post-its and cardboard, to 3D models, to polystyrene milled out in the correct measurements and designs. The levers were developed from their original shapes through iterations of feedback from focus groups with experienced operators. They explained in detail the issues they had with the levers available today and were given the possibility to give direct input to the industrial designers on which changes they wanted. The designers then presented 3D drawings of the input devices, where the operators could vote and comment on the suggestions they preferred. At the next focus group meeting, physical clay models of the input devices were presented. The operators could then experience their feedback and shape the devices according to their preference. To counteract the effect where only a few could influence the design, the developers took the final 3D-printed models on tour on board vessels to ask on board crew their opinions. In addition, general opinions were gathered from inexperienced users to gain feedback from a different perspective.

A similar prototyping procedure was carried out for the development of the new operator chair. Looking to other industries such as the automotive industry gave inspiration to the design. The seats in sports cars often has a midfield of alcantara, a synthetic, durable and breathable fabric. This gave friction when seated, in addition to removing dampness when being seated for a long time. It is all connected to removing the little issues that might seem like a trifle but could potentially be the drop that could make the glass spill over in a safety-critical situation. Not alone, but together with all the other little trifles. During observations, it was discovered that the operator several times had to almost lean out of the chair to get a glimpse of the things going on behind the chair. By changing the shape of the chair to a pear-shaped/drop shape backrest, the chair had a narrower headrest than traditional operator's chairs, increasing the operator's field of vision (Figure 9.3). The operator could now fit more information into the corner of the eye without moving his/her head.

The final prototype was presented at a large maritime fair as a market ready concept. The different projects were assembled together, presenting consoles, levers/input devices and a new operator's chair. They were sold to a simulator centre where they are utilized today during training of maritime personnel.

TESTING

The first Unified Bridge contract was signed with a Norwegian ship owner, where the concept was to be delivered to a platform supply vessel. Several test iterations had been carried out throughout the previous stages of the process and the concept had been displayed and demonstrated for customers and stakeholders at a national maritime fair. However, for the first customer delivery the ship bridge environment was assembled as a full-scale model, where all technical and ergonomic requirements were tested; in addition, the concept was run through a factory acceptance test (FAT), which is a standard procedure when new products are to be installed on a vessel. During a FAT, external testers evaluate the equipment during and after the assembly process by verifying that it is built and operating in accordance with design specifications. When accepted, the ship bridge environment was disassembled and shipped to the shipyard for instalment on board and where the final test was carried out by the classification society, giving the final certificate for the vessel to be allowed to sail.

STANDARDS AND REGULATIONS

When being innovative in the maritime industry and presenting solutions that have not been evaluated earlier to stakeholders, the solutions on board needed acceptance in the classification regulations given by the classification society. DNV-GL's NAUT-OSV standard (DNV-GL, 2012) was the main working document that the development of the innovative solutions needed to adhere to for the vessel to be allowed to sail on its maiden voyage. The NAUT-OSV classification regulation (DNV-GL, 2012) contains some sections that reviews topics such as the design of workplace on board. This section specifies the requirements for bridge design, including field of vision, wheelhouse arrangement, workstation configuration and location of equipment within workstations. The regulations can in many cases be widely interpreted and include the minimum of necessary requirements to take care of the human element. Methodologies or processes on how to do this, such as utilizing design thinking or the user centred design process, are not a part of the regulations. To comply with the rules and regulations, the solutions presented were in many cases far from the traditional solutions. One example that was costly both financially and in time was the need to clean up the clutter on the worksurfaces. To be able to fulfil this requirement, equipment integration was the keyword to make this happen. The possibility to introduce new solutions that could leave only the operation critical input devices close to the operator meant that other equipment had to give way. Equipment that was needed for the operation, however not critical to the manoeuvring of the vessel, such as windscreen wipers, lantern control, phone list (often a laminated piece of paper), de-icing system bridge light control, etc., was categorized as "support systems" (Figure 9.4). These systems were often delivered from a range of different suppliers who had no intention to harmonize their look and feel with other suppliers. The panels are often large with a lot of specialized functions rarely utilized, which contributed to cluttering the work surfaces. The solution lies in integration. By integrating the interfaces (signals) of the different suppliers into

FIGURE 9.4   Remote control display for support equipment (orange circle) and illustrating increased field of vision to the aft deck.

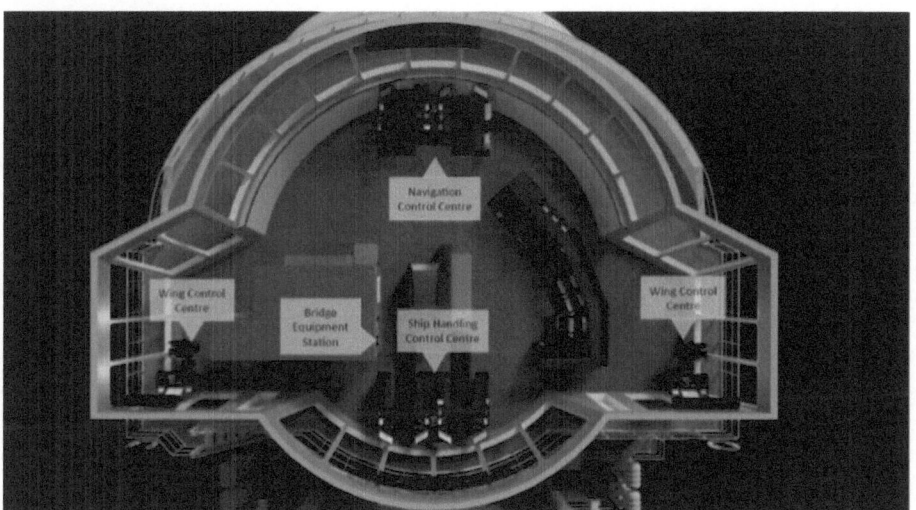

FIGURE 9.5 Unified Bridge layout.

one product, remote controlling the different systems through a 10″ touch panel mounted next to the operator becomes possible. The operator has all the necessary functionality available at his/her fingertip, and it would be possible to save important space by moving the actual panels to an equipment station further away from the workstation. This would bring the wanted effect of a clean and clutter-free console with only the necessary panels and devices represented in immediate proximity to the operator (Figure 9.5).

When integrating different systems, it cleans the visual clutter, but introduces a new level of complexity "under the bonnet". The classification society viewed this complexity as a risk as the high level of integration was new and not clearly explained in the regulations leading to the classification society naturally being reluctant to approve the new product. This caused an expensive and lengthy process to obtain the approval. When being on the frontlines of innovation in a conservative industry, one must include the possible risk of higher cost to pave the road for others. The example above was one of many similar examples, which again would lead to a higher financial entrance and an increased project timeline.

Throughout the development of the Unified Bridge, the company had a permanent contact in classification society to make sure that the dialogue with the classification and surveyor part of classification society went smoother. This was a benefit to both the classification society and to the supplier of maritime equipment. Although a benefit, that does not necessarily mean it is a clear path to receiving the necessary approvals. The classification regulations are as comparable to other law and regulations documents, where the content can be interpreted in different directions. This means that consultancy hours purchased to make sure products were developed according to regulations does not guarantee the certificates as the surveyors approving the vessel may interpret the regulations differently than the consultants.

EXPERIENCES, BENEFITS AND CHALLENGES

Incorporating design thinking and an user-centred design process in the development of a ship bridge environment has been challenging and costly but holds benefits that the operators find highly advantageous. A study done by Danielsen et al. (2019) investigates the ship bridge environment on board two platform supply vessels with the Unified Bridge installed. The vessels had been in operation for 5 years with low turnover on the bridge crew and their feedback were undoubtedly positive. The crew described the Unified Bridge as being well arranged and an user-friendly environment to work in. These findings give strong indications that the human-centred design process behind the development of this ship bride seems to have been able to accommodate many of the end-user needs. The seafarers claim that the design makes sense to them and is in line with their practices of work. Although there were some points of improvement, these improvements were not significant in comparison with earlier challenges with different suppliers of ship bridge environments. However, the crew did spark a concern regarding the crew's autonomy as integrated bridges, such as the Unified Bridge, may increase supervision and control from shore.

In addition to a few design issues, the development process initiated some challenges:

– The industry is conservative with low adaptability.
– The industry has low level of standardization of equipment placement.
– In the supply chain, the shipyard is a strong stakeholder with a strong cost focus.
– The classification rules and regulations did not follow the speed of innovation and compromises had to be made that impaired the quality of the user experience.
– High production costs of levers, consoles, chair and displays due to a low initial volume.
– High expectations from customers who wanted a conventional ship bridge with a high level of tailoring according to personal preference or due to that they had "always done it like that". This counteracts the purpose of the Unified Bridge, where one of the requirements was to simplify and standardize. One delivered vessel actually ended up with more displays and monitors than systems, as strong voices from the ship owner always wanted to have everything available, which was not the intention with the concept. No operator can monitor everything at the same time. This exemplifies the conservative mindset often met in the industry, which challenges the intention behind innovation and novel concept development.

The experiences gained from the development of the Unified Bridge point out some focus areas:

– There must be a strong and well-organized project group and organization around the concept development to keep the project on track. This was a success factor for the Unified Bridge project.

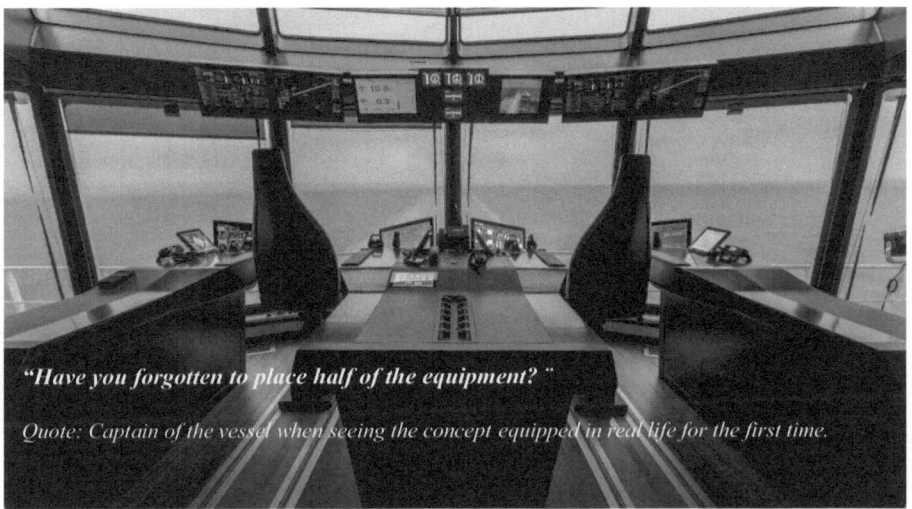

FIGURE 9.6   The Unified Bridge aft bridge workstation.

- Have well-developed connections into classification and regulation societies to be able to push the development of the regulations in an innovative direction.
- When moving from the prototyping phase (the first two vessels delivered was considered prototypes) to the phase of industrialization, compromises concerning design and user experience are likely to occur as the elements can be too costly to produce/industrialize.
- Make sure there is a strong organization within all fields (including marketing, engineering and management) around the finished delivery ready concept. This organization must work in a long-term perspective on changing the established and conservative culture concerning the design and equipping of ship bridge environments.

The overall experience is positive, especially after the end-user perspectives were collected after 5 years (Danielsen et al., 2019). There is however a high potential that conservatism and an engineering-focused culture will slow down the design thinking and user-centred design process. This can be counteracted by an organization that truly focuses, from top management and down throughout the organization, on design thinking and the user-centred design process and establishes this in its own quality management procedures (Figure 9.6).

REFERENCES

Allianz Global Corporate & Specialty. (2012). *Safety & Shipping 1912–2012*. From Titanic to Costa Concordia, Munich, Germany.

Bjørneseth, F.B., Dunlop, M.D., Hornecker. E. (2012). Assessing the effectiveness of direct gesture interaction for a safety critical maritime application. *International Journal of Human-Computer Studies*, 70(10), 729–745.

Bjørneseth, F.B., Dunlop, M/D., Clarke, L. (2014). Towards and understanding of operator focus using eye.tracking in safety critical maritime settings. *Human Factors in Ship Design and Operation, Conference*, Royal Institution of Naval Architects, RINA.

Brown, T. (2008). Design thinking. *Harvard Business Review*, 86(6), 84.

Danielsen, B.E., Bjørneseth. F.B., Vik, B. (2019). Chasing the end-user perspective in bridge design. *Proceedings of Ergoship 2019*. HVL, campus Haugesund.

Dekker, S. (2004). *Ten Questions about Human Error: A New View of Human Factors and System Safety*. CRC Press.

Dhillon, B. S. (2007). *Human Reliability and Error in Transportation Systems*. Springer Science & Business Media.

DNV-GL (2012). *Rules for Classification of Ships, Part 6 Chapter 20 Nautical Safety – Offshore Service Vessels*. DNV-GL, Høvik, Norway.

Holzinger, A., et al. (2012). On complexity reduction of user interfaces for safety-critical systems. *International Conference on Availability, Security, and Reliability, IFIP Cross Domain Conference*, LNCS 7465 (2012), 108–122.

Howell, W. C., Cooke, N. J. (1989). Training the human information processor: A review of cognitive models. In I. L. Goldstein (ed.), *Frontiers of Industrial and Organizational Psychology, The Jossey-Bass Management Series and The Jossey-Bass Social and Behavioral Science Series. Training and Development in Organizations* (pp. 121–182). Jossey-Bass.

IEC 61511. (2018). *Functional safety - Safety instrumented systems for the process industry sector*.

ISO 6385 (2004). *Ergonomic principles in the design of work systems*. ISO, Geneva.

ISO 9241-210 (2019). *Ergonomics of human-system interaction—Part 210: Human-centred design for interactive systems*. ISO, Geneva.

ISO 11064-1 (2000). *Ergonomic design of control centres—Part 1: Principles for the design of control centres*. ISO, Geneva

ISO 11064-2 (2000). *Ergonomic design of control centres—Part 2: Principles for the arrangement of control suites*. ISO, Geneva.

ISO 11064-3 (1999). *Ergonomic design of control centres—Part 3: Control room layout*.

ISO 11064-4 (2013). *Ergonomic design of control centres—Part 4: Layout and dimensions of workstations*. ISO, Geneva.

ISO 11064-5 (2008). *Ergonomic design of control centres—Part 5: Displays and controls*. ISO, Geneva.

ISO 11064-6 (2005). *Ergonomic design of control centres—Part 6: Environmental requirements for control centres*. ISO, Geneva.

ISO 11064-7 (2006). *Ergonomic design of control centres—Part 7: Principles for the evaluation of control centres*. ISO, Geneva.

Kilskar, S.S., Danielsen, B.E., Johnsen, S.O. (2019). Sensemaking in critical situations and in relation to resilience—A review. *ASCE-ASME Journal of Risk and Uncertainty in Engineering Systems, Part B: Mechanical Engineering*. doi:https://doi.org/10.1115/1.4044789

Kirwan, B., Ainsworth, L.K. (1992). *A Guide to Task Analysis, the Task Analysis Working Group*. CRC Press.

Marine, L. (2014). Task analysis: The key UX design step everyone skips. *Search Engine Watch*. Retrieved from: http://searchenginewatch.com/sew/how-to/2336547/task-analysis-the-key-ux-design-step-everyone-skips, 08.06.2020.

Nocera, D., et al. (2007). A random glance at the flight deck: Pilots' scanning strategies and the real-time assessment of mental workload. *Journal of Cognitive Engineering and Decision Making,* 1, 271–285.

Weick, K. E. (1988). Enacted sensemaking in crisis situations. *Journal of Management Studies,* 25(4), 305–317.

Weick, K. (1993). The collapse of sensemaking in organizations: The Mann Gulch disaster. *Administrative Science Quarterly,* 38, 628–652.

10 Supporting Consistent Design and Sensemaking across Ship Bridge Equipment through Open Innovation

S. C. Mallam
University of South-Eastern Norway

K. Nordby
Oslo School of Architecture and Design

CONTENTS

INTRODUCTION

A ship's bridge is a complex work environment that is typically outfitted with a range of equipment supplied from numerous different vendors (Lützhöft & Vu, 2018). Multivendor ship bridge systems (MBS) that merge independent equipment from different vendors can create disparities for navigation crew once all necessary equipment is installed into a single work environment (Nordby, Mallam & Lützhöft, 2019). Navigators must engage with different pieces of equipment located on the bridge of a ship throughout their watch shift to enable successful planning and execution of operations. Thus, the very nature of MBS create circumstances where navigators must interact with equipment having multiple design languages across the many physical and digital inputs and outputs of a bridge (see Figure 10.1). For example, different software systems, even supplied by the same vendor, may have different screen layouts, menu structures, colour combinations, iconography, font style and sizes that vary across equipment. A simple analogy to the inconsistencies of typical MBS are the differences found between using contemporary Mac and Windows PC operating systems. Users must adapt to the individual systems and functionalities in order to successfully complete desired tasks. Poor design increases cognitive demands for users (Woods & Patterson, 2000) and can have negative consequences, particularly in the safety-critical context of operations at sea (Lee & Sanquist, 2000; Mallam, Lundh & MacKinnon, 2015). Poor Graphical User Interface (GUI) design has shown to have negative implications on navigation operations, increasing the potential for making errors, hiding critical information and contributing to accidents and deaths at sea (Kataria, Praetorius, Schröder-Hinrichs & Baldauf, 2015; Mallam, Nordby, Johnsen & Bjørneseth, 2020).

This chapter outlines how current conditions in the maritime industry, regulatory scope and ship procurement processes that create an ecosystem where poor design

FIGURE 10.1   Example of an multivendor ship bridge systems (MBS) centre console of a contemporary patrol vessel.

solutions can become more readily introduced and embedded in bridge design, and thus affect seafarers and their sensemaking during operations. The OpenBridge Design Guideline is presented, which addresses these deficiencies by developing interface guidelines based on web technologies, user-centred design principles and component libraries. We detail the background and philosophy of OpenBridge, including how it fills gaps in current industry processes and regulatory frameworks, and strive for a more integrated approach through open innovation. Finally, we present a case study applying the OpenBridge Design Guideline to the development of an ECDIS (Electronic Chart Display and Information System) and discuss its contributions for enhancing sensemaking in ship operations.

CURRENT SHIP PROCUREMENT PROCESSES INTRODUCE INHERENT DESIGN DEFICIENCIES

There are a range of influencing factors for why inconsistent bridge systems are prevalent across the maritime industry. The typical contemporary procurement process for ship bridges have different systems provided by multiple vendors; however, this is not in and of itself the reason for such drastic design inconsistencies across bridge systems. The design and construction of ships are large-scale projects that typically take years from initial concept to a constructed vessel sailing in water (Eyres & Bruce, 2012; Veenstra & Ludema, 2006). Shipbuilding processes are often split between numerous stakeholders and geographical locations (Stopford, 2009; Österman, Ljung & Lützhöft, 2009). Ultimately, ship owners and investors generally focus on big picture issues of a ship's construction and specifications, such as cargo carrying capacity, speed, versatility and efficiency, as opposed to detailed design of the working environment (Eyres & Bruce, 2012).

Typical ship design processes are inherently engineering centric (e.g. ship design spiral [Evans, 1959]) and normally do not include the perspectives and knowledge of the seafarers themselves, or operational demands within design cycles (de Vries, Costa, Hogström & Mallam, 2017). Furthermore, those in charge of maritime equipment design generally do not have an understanding of operations or seafarer demands, leading to poor design choices (Chauvin, Le Bouar & Renault, 2008; United States Fleet Forces Command, 2017). A technology-centric implementation of bridge systems can lead to designs that do not support the users or sensemaking in operations (Johnsen, Kilskar & Danielsen, 2019). Without expert user knowledge integrated throughout design development, a disconnect can emerge between the final ship design, including its onboard equipment, and how the crew use the ship to accomplish their tasks (Mallam, Lundh & MacKinnon, 2015). This can lead to suboptimal and unsafe working practices, increasing the likelihood of errors and accidents.

Current bridge mandatory design regulations and non-mandatory design guidelines also fail to adequately support design consistency across bridge equipment (Mallam & Nordby, 2018). The International Maritime Organization's *International Convention for the Safety of Life at Sea* (SOLAS Convention) provides predominantly goal-based objectives for bridge work environment design (e.g. SOLAS chapter V/15) (International Maritime Organization, 2009). Additional guidance notes and references are provided in a relatively more prescriptive approach to bridge design (e.g. International Electrotechnical Commission, 2008; International Maritime Organization, 2000, 2004; International Organization for Standardization, 2007). However, in totality, detailed aspects of digital interface design guidance are still lacking. Efforts have been made by regulators and industry stakeholders to standardize aspects of GUI across the industry, such as symbols (International Maritime Organization, 2019a, 2019b) and for new technology development and deployment, such as the introduction of ECDIS (International Maritime Organization, 2017). Even guidelines from supporting classification bodies and regulatory authorities fail to capture necessary design guidance, not only for the bridge but for other operational areas of a ship, including the engine department (Mallam & Lundh, 2013).

As equipment and operations increasingly digitize, there is an opportunity to create design guidelines that support design consistency across all systems for multivendor and integrated bridges. Creating design consistency across equipment in the bridge will lead to positive outcomes in navigational operations and support-enhanced sensemaking of seafarers. This cannot be captured in regulations or design guidance alone. Rather, it must also be integrated with practical and usable processes that enable the application of knowledge and design guidance throughout ship and bridge development. This must engage the identified industry stakeholders and ultimately add value to the process and its outcomes.

SUPPORTING SENSEMAKING THROUGH DESIGN

Sensemaking is the process of turning circumstances into a situational understanding in which we can react upon (Weick, Stutcliffe & Obstfeld, 2005; Taylor & Van Every, 2000). By emphasizing sensemaking in the design process we aim to rationalize the user's actions by understanding how cognitive activities are employed to experience meaning in an interaction situation (Weick, 1993). These activities involve perceiving and interpreting sensory data, formulating and using information, and managing attention (Bartscherer & Coover, 2011). Although complex information spaces, such as a ship's bridge, continue to expand extensively as technology and automation develops, human capabilities and information processing struggle to adapt and successfully manage the complexity of operational systems (Bainbridge, 1983). According to Chia (2000), conception is shaped through "differentiating, fixing, naming, labelling, classifying and relating" sense-impressions systematically to construct our experience of social reality. Therefore, digital sensemaking requires mediation in order to meet the user's needs by labelling and categorizing content and choices in user interfaces to balance the streaming of experience (Weick, Sutcliffe & Obstfeld, 2005). Consistent user interface design is a strategy for digital sensemaking that not only supports easier interpretation, improved usability and productivity but also enhances the possibility to transfer skills across systems because users are able to predict how interfaces will look and function due to familiarity and experience (Nielsen, 2002). This makes the user more "fluent" when operating known systems and even new systems on first encounter. This is particularly relevant for many sectors of the maritime domain due to the nature and variability in how seafarer employment is organized. Although some seafarers may work within a single company, or even a single ship for years, there is a segment of the population that are employed in a relatively transient manner. Seafarers may work on different types of contracts and sign on/off different ships and ship types (whether within a single company, across multiple companies or across industry sectors). Work periods at sea may be organized by different durations (e.g. days, weeks or months at a time) and across short- or long-term work contracts (e.g. a single trip at sea for a specific time period vs. multi-trip contracts and more permanent/reoccurring employment opportunities). Thus, seafarers moving between different ships require a high degree of adaptability to potentially unfamiliar and inconsistent work environments, operational systems and GUI (in addition to different work colleagues, cultures, policies, procedures, routes, etc.).

Design consistency is critical in supporting sensemaking during operations, as sensemaking requires interpretation and action, enacting a more ordered environment from which further cues can be drawn to create a shared understanding (Kilskar, Danielsen & Johnsen, 2019). The performance of users is directly influenced by design, and poor solutions can introduce negative effects for the user and the system (Dul, Bruder, Buckle, Carayon, Falzon, Marras, Wilson & van der Doelen, 2012). In order to reduce usability problems and increase quality of interactions, design should provide affordances, or cues, that users can recognize, make decisions from and then implement correct actions in order to fulfil a desired task and outcome (Pucillo & Cascini, 2014). This is true of very simple interactions, from correctly interpreting and turning on a faucet for desired water temperature and rate of flow (e.g. "which handle for warm water?", "What angle or rotation for optimal flow?") or opening an unfamiliar door ("Do I push or pull? Will it open automatically... from the center?") to complex safety-critical interfaces and working environments, such as airplane cockpits or central control rooms of power plants. Similarly, in the maritime context of a ship's bridge, the design quality and design consistency across interfaces and interaction devices between navigator and equipment are critical for reducing user errors and increasing productivity. Design should impart relevant information and effectively communicate how to achieve the intended task or tasks to its users. Interaction with an object, interface or system, whether physical or digital, should easily communicate relevant information to the user that is interpretable and enables optimal decision-making for their intended actions.

Good design enables sensemaking in navigators by supporting and guiding expectations of a system, its current operational status and future requirements. This is becoming increasingly relevant with digitized interfaces and the characteristics of information on screens. In particularly, how information is organized in digital menus and sub-menus has implications for how users navigate through a software program and the ease of access to relevant and desired information. As the investigation of a recent United States Navy accident found (*USS John S. McCain collision with Alnic MC in 2017*), the design of digital systems and how information is presented can mislead or inhibit users from correctly interpreting a situation and operational status (National Transportation Safety Board, 2017).

SENSEMAKING FOR CURRENT AND FUTURE SHIPPING OPERATIONS

Like most industries, throughout the 20th and 21st centuries automation has led to both higher operational productivity, which, in turn, also reduced the number of required personnel. Ship operations are increasingly automated and centralized, requiring less direct implementation of tasks but increased monitoring and troubleshooting of automated processes. As the proliferation of digitalization of front–end interaction and display devices increases, the maritime domain finds itself at crossroads for future operational systems. Current design guidelines fail to adequately address detailed design aspects of digitized systems and GUI for maritime systems, while the most robust, well-established design guidance focuses on the physical aspects of bridge work environments, console construction and layout (Mallam & Nordby, 2018). This presents a timely opportunity to develop and implement common frameworks for digital GUI systems that are currently lacking within the maritime domain.

There are additional spin-off benefits for consistency of control stations (i.e. the bridge, engine control room, shore-side control centre) for the maritime domain as a whole, including shipping companies, equipment manufacturers, policymakers, seafarers and other relevant stakeholders. For example, classification bodies and regulatory authorities can establish consistency in bridge systems across fleets or industry sectors, creating more streamlined inspection and approval processes. Increased consistency would also allow seafarers to switch between ships while maintaining mastery between interfaces, workstations and work environments. This would have additional benefit of reducing training and familiarization periods between seafarers and equipment. As ships are large monetary investments and have operational lifecycles at sea spanning several decades, retrofitting of a ship structure and its systems are required periodically. Digital systems and GUI are more flexible to develop, cheaper and faster to upgrade, alter and improve in comparison to hardwired and analogue systems (e.g. analogue control console vs. software and screens). Digital systems have a lower threshold for both development and implementation, making updates potentially more economically favourable, as well as more effective in creating better design solutions, not only at the new-build stage but also for retrofitting across a ship's lifecycle.

Furthermore, interest is increasing in unmanned and autonomous vessels that are remotely controlled and monitored. Although still in relatively early phases, MASS (Maritime Autonomous Surface Ships) are developing rapidly in research and development phases throughout the world (International Maritime Organization, 2018; Pribyl & Weigel, 2018; World Maritime University, 2019). Remote control and/or monitoring of unmanned ships at sea are currently being developed through differing shore-side control centre concepts. However, both "traditional" operations of contemporary navigational tasks at sea and future configurations of shore-side control and monitoring will continue to require interactions between people and technology through GUI in some capacity (Kim & Mallam, 2020). The design of equipment and interactions between technology and people must support sensemaking and is perhaps even more critical in the age of distributed command and control of ships at sea where operators and monitors of remote structures will need to have all relevant information to enable optimal sensemaking throughout operations.

TOWARDS IMPROVING DESIGN CONSISTENCY

Consistent design refers to a common set of design principles that are shared by different systems (Nielsen, 2002). Typical examples of such principles are palette definitions or user interface components, such as buttons. Design consistency leads to a range of benefits, including increased usability, ease of learning and reduced errors (Nielsen, 2002). These benefits increase with the number of systems that apply the same design principles. Although the benefits of consistent design are well known, current maritime design approaches and regulations have not managed to realize the full potential of consistent design. Overall, we argue that each of the individual approaches above have shortcomings and will likely not lead to design consistency alone. Instead, there is a need for a combined approach. Below we describe four design approaches and analyse how they impact the push towards design consistency.

PRESCRIPTIVE RULES

Prescriptive rules and guidelines specify in detail how a product or system should be designed. This may include detailed design guidance of elements including, font sizes, icons characteristics and specific functions applications must deliver. In the maritime domain, there are both mandatory prescriptive rules and non-mandatory prescriptive guidelines for design. A limitation of prescriptive rules is that they are typically established from prior experiences and thus may have inherent biases in their scope and content. Prescriptive rules related to engineering and technology-related fields are also challenging due to the rapid evolution of contemporary technological advancements. By the time the need for prescriptive rules are identified, developed and released, the applications the rules were originally intended for may have less relevancy or suitability. Currently, there does not exist prescriptive regulation in the maritime sector that defines user interfaces at a level that may lead to digital design consistency. The current rules manage some aspects of central applications, such as ECDIS; however, none offer detailed specifications of user interface design.

GOAL-BASED RULES

Goal-based rules and guidelines define what a system should achieve, instead of how it should be designed. These guidelines are useful in establishing high-level goals (e.g. SOLAS chapter V/15 [International Maritime Organization, 2009]); however, it is difficult to achieve consistent design using goal-based rules, since a goal can be achieved in many different ways, through many different solutions. Although it is possible to mandate that a system should be designed consistently, it is difficult to realize this in multivendor ecosystems, where many systems are developed independently by independent actors. As a result, a goal-based rule approach for realizing consistent design may work within a single company or group of collaborating companies, but it is likely to fail when systems are designed and integrated by several competing companies.

MANUFACTURER CONTROLLED CONSISTENCY

Typically, larger vendors who control the design of an entire ship's bridge deliver the most consistently designed systems in the maritime sector. These companies may achieve consistency across a single ship's bridge or several ship bridges by controlling a large portion of the design process and its outcomes. However, this approach is limited to a company's own available systems and product offerings, meaning that ships using equipment from competing manufacturers will no necessarily share the same design principles. Because a single manufacturer rarely delivers all equipment of a single workplace, installed third-party equipment will still create inconsistencies within a bridge work environment.

DESIGN TRANSFER FROM OTHER FIELDS

In recent years, design principles from the web and mobile industries have begun to enter the maritime domain driven by low-cost, high-quality design guidelines and technologies. These tools are slowly entering the maritime domain across systems in the form of user interface components and design patterns. Over time, this may contribute to system consistency, independently of system vendor. However, there are challenges using such components since their design guidelines were not originally intended for a maritime context and lack specific maritime-related content.

THE OPENBRIDGE CONCEPT

OpenBridge is an initiative that began with the aim to improve the usability of user interfaces in maritime workplaces through collaboration between maritime system developers (Nordby, Frydenberg & Fauske, 2018). The consortium is developing an open innovation approach to maritime workplace design by offering common design guidelines, implementation methods and approval systems. The OpenBridge vision is to simplify both the design and implementation of maritime systems in order to reduce development costs and improve the usability of product offerings.

OpenBridge is based on precedence from the web industry, where freely available frameworks for user interface development have been widely adapted, resulting in reduced costs and improved usability. In order for success, any new design system initiative must be desirable for the commercial interests of its users and the industry at large. Consequently, ensuring the design system is attractive from an economic perspective is of high priority. This includes developing reusable design patterns, tools supporting both design and implementation processes, and reusable code. Reusable code has a dual function: first, it reduces the cost of technical implementation. Second, it reduces the need for developers to interpret the design guidelines when creating new interfaces, since many of the components described in the guideline are already created and catalogued.

THE OPENBRIDGE APPROACH TO CONSISTENCY

Navigators are constantly interacting with internal bridge equipment (e.g. electronic charts, RADAR, conning display, communication channels), other onboard crew and departments of the ship (e.g. engine, deck crew) and the external environment (e.g. weather conditions, geography, sea traffic, port authorities, vessel traffic services). Individual bridge equipment itself comprises a hierarchy of elements, which in combination create the interaction experience of its user interfaces, including hardware and software components. Each component of the bridge affects navigators' sense-making and the totality of their user experience. Thus, the design of equipment can be optimized for each of these components and ultimately extended to MBS integration of other equipment into a single working environment.

The OpenBridge Design Guideline is a component-oriented, prescriptive framework that seeks compatibility with existing prescriptive and goal-based regulations. A core concept is that the OpenBridge Design Guideline is divided into a series of independent components described through a user interface architecture (Nordby, Gernez & Mallam, 2019). The component structure draws upon the design guidelines as a hierarchy of elements that together make up OpenBridge-compatible user interfaces. The user interface architecture includes components, such as palettes, typography and layout patterns. Each element defined in the guideline follows current mandatory regulations to ensure that vendors following the OpenBridge Design Guideline also adhere to industry standards.

When designing a new system, a developer may use every element from the OpenBridge guideline or just a selection of them. In our perspective, designing for consistency should be a long-term and gradual process. It is not a binary output where a system is deemed "consistent" or "not consistent" but rather how many and which consistent attributes a system possesses. In this perspective, we suggest that two different applications are marginally more consistent if they share the same palettes, even though the rest of the applications may be completely inconsistent.

Importantly, OpenBridge focuses on generic user interface components and some core functions requiring consistency. Examples include a standard navigation menu and alarm display that are shared across all bridge systems. However, a degree of freedom is built into the OpenBridge Design Guideline in how to design the core functions of an application. We argue these parts of a system are better served by goal-based guidelines for generally "acceptable" usability levels. Thus, developing OpenBridge is a matter of balancing prescriptive generic user interface design with goal-based functions. Different industry stakeholders will have varying needs and apply the guideline partially or in its entirety (see Table 10.1). For example, integrators and system developers that are dependent on each other have an immediate benefit of fully implementing the OpenBridge Design Guideline. OpenBridge-compatible equipment can be sold to more ships without changes, and integrators can choose between more equipment that is out-of-the-box ready to use without further need for adaptation.

ECDIS CASE STUDY: APPLYING OPENBRIDGE FOR ENHANCED SENSEMAKING

THE DESIGN PROCESS

In the following sections, we present how the OpenBridge Design Guideline was used to design a version of an ECDIS user interface. The ECDIS design was developed in an iterative process that included analysis of prior art and in collaboration with domain experts, including developers, equipment manufacturers and maritime

TABLE 10.1

Examples of Different Types of Stakeholders That Have Different Needs in Applying the OpenBridge Design Guideline

Stakeholder Type	OpenBridge Implementation
Independent integrator	They can pick and choose any of the compliance elements
Independent equipment manufacturer not concerned about integration	They can pick and choose any of the compliance elements
Integrator that wants to integrate OpenBridge equipment	Need to follow rules up to integration system structure
Equipment manufacturer that wants to adapt to OpenBridge integrators	Need to follow rules up to applications structure
Integrator that wants to maintain a non-OpenBridge compliant look but integrate third party equipment	Need to follow minimal integration conventions

personnel. Through this initial process we have uncovered several challenges across different areas of the design guideline that provides feedback for further refinement and improvement for its content and application in future design processes. It is important to note that the example presented has not been through an approval process and must be considered as an early-phase development sketch. However, this case study sheds light on how different levels of consistency offered by OpenBridge are applied and how it affects users' sensemaking at each level.

PALETTES

OpenBridge has four palette settings for maritime user interfaces: bright, day, dusk and night (see Figure 10.2). The user interface components in OpenBridge are built upon these palettes, meaning that developers of maritime applications do not need to invest resources on colour palette or colour contrast development challenges. The palette system affects sensemaking in three ways. First, it is designed to offer optimized contrast and readability, which supports general sensemaking. For example, the "day" and "bright" palettes offer Level AAA contrast relative to Web Content Accessibility Guidelines (WCAG). Second, the palette uses its colour scheme to create meaning. For example, fonts, numerical data and interface sections are coded with consistent colours to improve the user's ability to decipher the content and meaning within the user interface. This has the potential to improve sensemaking of a single user interface, and when applied to several interfaces, it will also contribute to improving users' sensemaking across interfaces. Third, in ship bridges, different palettes can create light pollution problems for users. A common palette across systems will help reduce these inconsistencies. This has a particular advantage on the bridge during operations in the dark, where small differences in screen light emittance levels can impact the ocular adaption of seafarers on duty more so than in other conditions.

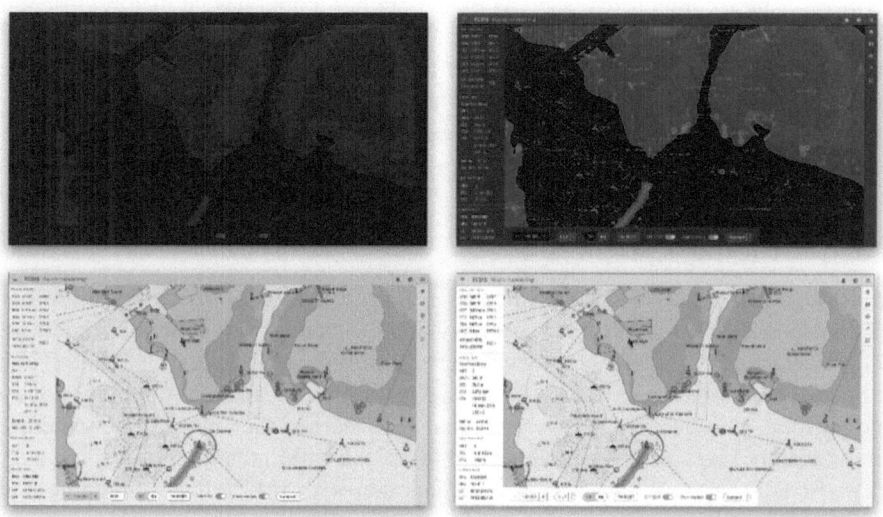

FIGURE 10.2 Examples of electronic chart display and information system (ECDIS) interfaces with different palettes.

TYPOGRAPHY

OpenBridge typography was established to have a limited set of text with distinct styles to represent specific types of information. Designers then apply the different text styles according to the specified functions in the interface, thus increasing design development efficiency and consistency by removing the need to create individual typography systems. Typography affects sensemaking in a similar way as palettes. First, OpenBridge has a limited choice of typesets, which requires designers to simplify how typography is used. Second, the typography is coded to help users navigate the interface, generating a clear separation between different elements (e.g. headings, labels and instrument text), thus supporting the users' ability to locate desired information within the interface and reduce screen clutter. Third, the meaning of the different typesets are standardized. Thus, typography provides design affordance, or cues, embedded within the typeset that users will learn, and maintains consistency across all other OpenBridge-compatible interfaces.

USER INTERFACE COMPONENTS

The OpenBridge Design Guideline has a selection of user interface components that can be used to build an interface. This ranges from simple components, such as buttons, to complex nested components, such as the application navigation menu. The design guideline also includes maritime-specific components, such as compass and rudder. The ECDIS user interface was created using ready-made components already established in the guideline. The user interface components support sensemaking in offering predictable, consistent information presentation and interactions. In previous user interfaces, simple components, such as toggle buttons, could have radically different designs across different equipment (Nordby, Mallam & Lützhöft, 2019). By standardizing a set of common components, it is easier for users to understand key functions. We argue that an additional effect for improved sensemaking is that more predictable user interface components allow users to free their cognitive resources on actual maritime operations rather than in learning and interpreting the operational systems.

GENERIC APPLICATIONS STRUCTURE

OpenBridge introduces a strict application structure that is governed by a top bar (see Figure 10.3). This structure organizes generic functions (i.e. functions that most applications offer, e.g. navigation menu, alerts, dimming and applications centre) in a predictable manner across systems. In the navigation menu, specific functions have a fixed position. This concept allows users that have learned the OpenBridge structure to immediately know where to find typical functions within the interface.

The ECDIS case was developed in conjunction with RADAR development and was part of an integrated navigation system. In order to navigate the system, we applied a model that separates between operations, applications and tasks. Operations are managed on a workstation level, where a user can select an operation that would control what applications are visible on the stations around them, such as conning,

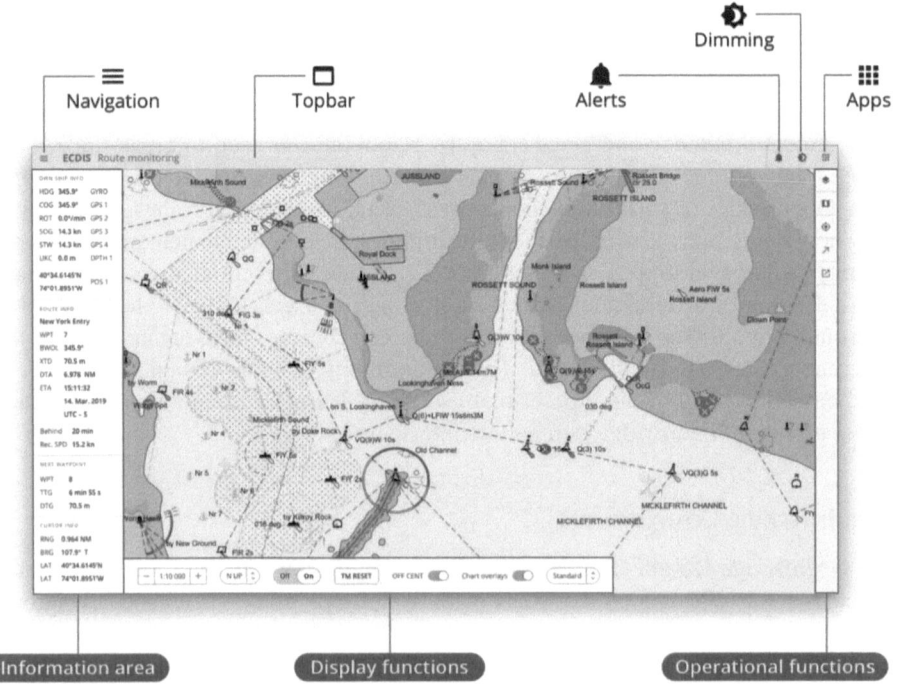

FIGURE 10.3   OpenBridge electronic chart display and information system (ECDIS) interface structure.

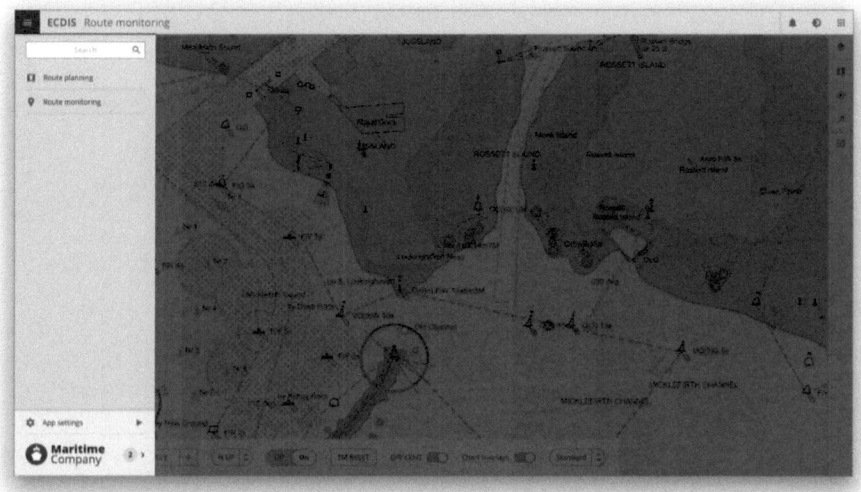

FIGURE 10.4   Application navigation menu (*left*).

ECDIS and RADAR. Each of these applications are managed through tasks, which are accessed through the application navigation menu (see Figure 10.4). The generic applications structure supports sensemaking in helping users to understand where typical applications functions are placed in an application. This is especially important for systems where users need to utilize multiple applications.

NAVIGATION AND INFORMATION PLACEMENT

After having applied the basic elements of the OpenBridge Design Guideline, the main functions of the ECDIS system were then implemented. OpenBridge does not offer any strict guidelines for how this should be performed, but rather establishes the structure around the navigation and information placement to sit within. For ECDIS design, the main functions in the display are separated into interactive functions (right side) and information elements (left side) (see Figure 10.5). The view control is positioned inside the map view to emphasize the connection between the map and the controls.

CONCLUSIONS

The ECDIS case reveals how the various types of elements defined in the OpenBridge Design Guideline supports users' sensemaking in individual applications, but more importantly, by making key attributes consistent with other applications. We also show how many aspects of an interface can be replaced by reusable components that are also shared across other interfaces. This feature has a dual effect: components may achieve improved usability while simultaneously saving development costs for

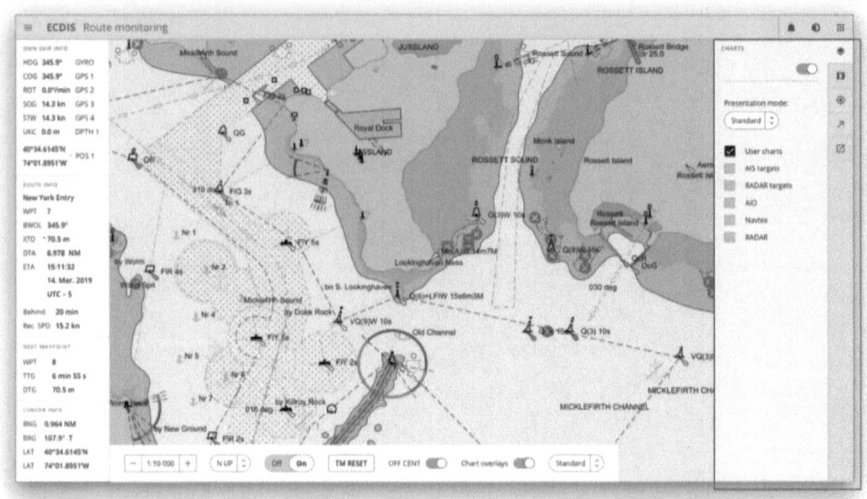

FIGURE 10.5 Electronic chart display and information system (ECDIS) main functions separated into interactive functions (*right side*) and information elements (*left side*).

manufacturers. This case also reveals how the prescriptive nature of the OpenBridge Design Guideline can be applied in conjunction with goal-based guidelines. The prescriptive guidance of OpenBridge fills current gaps for digital user interface design while complimenting the relatively higher-level general guidance currently in existence within the maritime domain. The OpenBridge Design Guideline is applicable and relevant not only for the maritime domain but can be expanded to other industry's digital interfaces and systems that require consistency. As components of the OpenBridge Design Guideline are generic, it is possible to generalize and apply the fundamental properties to non-maritime uses, expanding the relevancy and broader impact of the guideline for other industries facing similar challenges with digital transformation.

We argue for an open approach for innovation. The strength of an open innovation approach is through momentum and adoption: the value, impact and quality of OpenBridge and the OpenBridge Design Guideline increases as more actors use, test and contribute to its development. The OpenBridge Design Guideline is neither static nor definitive. Rather, it must evolve and change as the industry, technologies and operations evolve and change. As OpenBridge is increasingly applied to real-world applications and design cases the more robust and relevant it will become, ultimately improving design outcomes and sensemaking for end users.

REFERENCES

Bainbridge, L. (1983). Ironies of automation. *Automatica, 19*(6), 775–779. DOI: 10.1016/0005-1098 (83)90046-8

Bartscherer, T. & Coover, R. (2011). *Switching Codes—Thinking Through Digital Technology in the Humanities and the Arts*. Chicago, IL: University of Chicago Press.

de Vries, L., Costa, N., Hogström, P. & Mallam, S. (2017). Designing for safe operations—promoting a human-centred approach to complex vessel design. *Ships and Offshore Structures, 12*(8), 1016–1023. DOI:10.1080/17445302.2017.1302637

Chauvin, C., Le Bouar, G. & Renault, C. (2008). Integration of the human factor into the design and construction of fishing vessels. *Cognition, Technology & Work, 10*(1), 69–77. DOI:10.1007/s10111-007-0079-7

Chia, R. (2000). Discourse analysis as organizational analysis. *Organization, 7*(3), 513–518.

Dul, J., Bruder, R., Buckle, P., Carayon, P., Falzon, P., Marras, W. S., Wilson, J. R. & van der Doelen, B. (2012). A strategy for human factors/ergonomics—developing the discipline and profession. *Ergonomics, 55*(4), 377–395. DOI:10.1080/00140139.2012.661087

Evans, J. H. (1959). Basic design concepts. *Journal of the American Society of Naval Engineers, 71*(4), 671–678. DOI:10.1111/j.1559–3584.1959.tb01836.x

Eyres, D. J. & Bruce, G. J. (2012). *Ship Construction* (7th ed.). Oxford, UK: Butterworth-Heinemann.

International Electrotechnical Commission. (2008). *Corrigendum 1-Maritime Navigation and Radiocommunication Equipment and Systems—General Requirements—Methods of Testing and Required Test Results* (IEC 60945:202/COR1:2008). Geneva, Switzerland: International Electrotechnical Commission.

International Maritime Organization. (2000). *Guidelines on Ergonomics Criteria for Bridge Equipment and Layout* (MSC/Circ.982). London, UK: International Maritime Organization.

International Maritime Organization. (2004). *Performance Standards for the Presentation of Navigation-Related Information on Shipborne Navigational Displays* (Resolution MSC.191(79)). London, UK: International Maritime Organization.

International Maritime Organization. (2009). *International Convention for the Safety of Life at Sea* (5th ed.). London, UK: International Maritime Organization.

International Maritime Organization. (2017). *ECDIS – Guidance for Good Practice* (MSC.1/Circ.1503/Rev.1). London, UK: International Maritime Organization.

International Maritime Organization. (2018). *Working Group Report in 100th Session of IMO Maritime Safety Committee for the Regulatory Scoping Exercise for the Use of Maritime Autonomous Surface Ships (MASS)*. London, UK: International Maritime Organization.

International Maritime Organization. (2019a). *Guidelines for the Presentation of Navigational-Related Symbols, Terms and Abbreviations* (SN.1/Circ.243/Rev.2). London, UK: International Maritime Organization.

International Maritime Organization. (2019b). *Guidelines for the Standardization of User Interface Design for Navigation Equipment* (MSC.1/Circ.1609). London, UK: International Maritime Organization.

International Organization for Standardization. (2007). *Ships and Marine Technology—Ship's Bridge Layout and Associated Equipment—Requirements and Guidelines* (ISO 8468:2007). Geneva, Switzerland: International Organization for Standardization.

Johnsen, S. O., Kilskar, S. S. & Danielsen, B. E. (2019). Improvements in rules and regulations to support sensemaking in safety-critical maritime operations. In *Proceedings of ESREL 2019–29th International European Safety and Reliability Conference.*

Kataria, A., Praetorius, G., Schröder-Hinrichs, J. U. & Baldauf, M. (2015). Making the case for crew-centered design (CCD) in merchant shipping. In *Proceedings of the 19th Triennial Congress of the International Ergonomics Association*, August 9th–14th, 2015, Melbourne, Australia.

Kilskar, S. S., Danielsen, B. E. & Johnsen, S. O. (2019). Sensemaking in critical situations and in relation to resilience—a review. *ASCE-ASME Journal of Risk and Uncertainty in Engineering Systems, Part B—Mechanical Engineering.* DOI:10.1115/1.4044789

Kim, T. & Mallam, S. (2020). A Delphi-AHP study on STCW leadership competence in the age of autonomous maritime operations. *WMU Journal of Maritime Affairs, 19*(2), 163–181. DOI: 10.1007/s13437-020-00203-1

Lee, J. D. & Sanquist, T. F. (2000). Augmenting the operator function model with cognitive operations—assessing the cognitive demands of technological innovation in ship navigation. *IEEE Transactions on Systems, Man, and Cybernetics, Part A—Systems and Humans, 30*(3), 273–285.

Lützhöft, M. & Vu, V. D. (2018). *Design for safety.* In H. A. Oltedal & M. Lützhöft (Eds.), *Managing Maritime Safety* (pp. 106–140). Abingdon, NY: Routledge.

Mallam, S. C. & Lundh, M. (2013). Ship engine control room design—analysis of current human factors & ergonomics regulations & future directions. *Proceedings of the Human Factors & Ergonomics Society 57th Annual Meeting, 57*(1), 521–525. DOI:10.1177/1541931213571112

Mallam, S. C., Lundh, M. & MacKinnon, S. N. (2015). Integrating human factors & ergonomics in large-scale engineering projects—investigating a practical approach for ship design. *International Journal of Industrial Ergonomics, 50*, 62–72. DOI:10.1016/j.ergon.2015.09.007

Mallam, S. C. & Nordby, K. (2018). *Assessment of Current Maritime Bridge Design Regulations and Guidance.* Report Prepared for the OpenBridge Project, The Oslo School of Architecture and Design.

Mallam, S. C., Nordby, K., Johnsen, S. O. & Bjørneseth, F. B. (2020). The digitalization of navigation—examining the accident and aftermath of US navy destroyer John S. McCain. In *Proceedings of the Royal Institution of Naval Architects Damaged Ship V*, 55–63.

National Transportation Safety Board. (2017). *Collision between US Navy Destroyer John S McCain and Tanker Alnic MC. Singapore Strait 5 Miles Northeast of Horsburgh Lighthouse August 21, 2017* (Marine Accident Report, NTSB/MAR-19/01 PB2019-100970). Washington, DC: National Transportation Safety Board.

Nielsen, J. (2002). *Coordinating User Interfaces for Consistency.* San Francisco, CA: Morgan Kaufmann Publishers.

Nordby, K., Frydenberg, S. & Fauske, J. (2018). Demonstrating a maritime design system for realising consistent design of multi-vendor ship's bridges. In *Proceedings of the Royal Institution of Naval Architects Human Factors*, 28–36. ISBN:9781510883208

Nordby, K., Gernez, E. & Mallam, S. (2019). OpenBridge—designing for consistency across user interfaces in multi-vendor ship bridges. In *Proceedings of Ergoship 2019*, 60–68. ISBN:978-82-93677-04-8

Nordby, K., Mallam, S. C. & Lützhöft, M. (2019). Open user interface architecture for digital multivendor ship bridge systems. *WMU Journal of Maritime Affairs, 18*(2), 297–318. DOI:10.1007/s13437-019-00168-w

Österman, C., Ljung, M. & Lützhöft, M. (2009). Who cares and who pays? The stakeholders of maritime human factors. In *Proceedings of The Royal Institution of Naval Architects Conference on Human factors in Ship Design and Operation*, 69–76. ISBN:978-190504055-1

Pribyl, S. T. & Weigel, A. M. (2018). Autonomous vessels—how an emerging disruptive technology is poised to impact the maritime industry much sooner than anticipated. *The Journal of Robotics, Artificial Intelligence & Law, 1*(1), 17–25.

Pucillo, F. & Cascini, G. (2014). A framework for user experience, needs and affordances. *Design Studies, 35*(2), 160–179. DOI:10.1016/j.destud.2013.10.001

Stopford, M. (2009). *Maritime Economics* (3rd ed.). Oxon, UK: Routledge.

Taylor, J. R. & Van Every, E. J. (2000). *The Emergent Organization—Communication as Its Site and Surface.* Mahwah, NJ: Erlbaum.

United States Fleet Forces Command. (2017). *Comprehensive Review of Recent Surface Force Incidents.* Norfolk, VA: United States Fleet Forces Command.

Veenstra, A. W. & Ludema, M. W. (2006). The relationship between design and economic performance of ships. *Maritime Policy & Management, 33*(2), 159–171. DOI:10.1080/03088830600612880

Weick, K. (1993). The collapse of sensemaking in organizations—the Mann Gulch disaster. *Administrative Science Quarterly, 38*, 628–652.

Weick, K. E., Stutcliffe, K. M. & Obstfeld, D. (2005). Organizing and the process of sensemaking. *Organization Science, 16*(4), 327–451. DOI:10.1287/orsc.1050.0133

Woods, D. D. & Patterson, E. S. (2000). *How unexpected events produce an escalation of cognitive and coordinative demands.* In P. A. Hancock & P. A. Desmond (Eds.), *Stress Workload and Fatigue.* Hillsdale, NJ: Lawrence Erlbaum.

World Maritime University. (2019). *Transport 2040—Automation, Technology, Employment—The Future of Work.* Malmö, Sweden: World Maritime University. DOI:10.21677/itf.20190104

11 User-Centred Agile Development to Support Sensemaking

M. E. N. Begnum
NTNU Gjøvik

CONTENTS

User-centred design (UCD) is a fitting approach for supporting sensemaking, as the aim of UCD is to understand contextual user needs and use this as a basis to iteratively explore and design solutions. Similarly, agile software development uses iterations and rapid feedback to continuously explore, learn and improve user stories and their implementations. The user-centred and agile approaches represent mindsets of great influence in their respective fields of design and software development. In this chapter, we will briefly introduce the two approaches. We present five models that can be used to combine agile and user-centred development and discuss three common challenges in user-centred agile development (UCAD). Key take aways from the chapter are:

- The importance of exploring, understanding and framing the problem, in order to:
  - Solve the right challenge.
  - Propose solutions that fit the context of use.
  - Make sure user needs drives ideation – not the love of technology.
- The importance of user involvement and contextual testing as a team effort to:
  - Enable interdisciplinary decision-making and collaboration.
  - Enable continuous learning and improvement based on feedback.
  - Create solutions that work in real-life scenarios and contexts.

A BRIEF INTRODUCTION TO USER-CENTRED DESIGN

The aim of UCD is to create solutions that fit the needs of the users, in their contexts of use. Therefore, an explicit understanding of contextual user needs is key (e.g. Forlizzi & Battarbee, 2004; Pruitt & Grudin, 2003; Sengers, Boehner, David, & Kaye, 2005; Vetere et al., 2005). Users – not technology – are in focus throughout the design process (Hartson & Pyla, 2012; Rubin & Chisnell, 2008). As sensemaking is dependent on cues from key systems, training, physical environments and social contexts of use, a user-centred approach is recommended for designing technology that is to assist human sensemaking.

UCD is used synonymously with human-centred design (HCD). However, some argue there are slight differences between these terms – for example, that swapping "human" for "user" emphasizes that we are talking about real people with emotions and psychological preferences – not some abstract "user", and that a number of stakeholders may be impacted instead focusing in on a selected user group. As UCD extends usability and includes a holistic user experience (UX), user-centred work is often referred to as "UX work".

UCD Core Principles

The core principles of UCD are (ISO, 2019):

1. The design is based upon an explicit understanding of users, tasks and environments
2. Users are involved throughout design and development
3. The design is driven and refined by user-centred evaluation
4. The process is iterative
5. The design addresses the whole UX
6. The team includes multidisciplinary skills and perspectives

Gould and Lewis (1985) underline how early and continuous focus on users, empirical measurement of usage and iterative design are core UCD traits. Still, there is a wide range in the degree of user contact in UCD processes – from thinking about the user via different

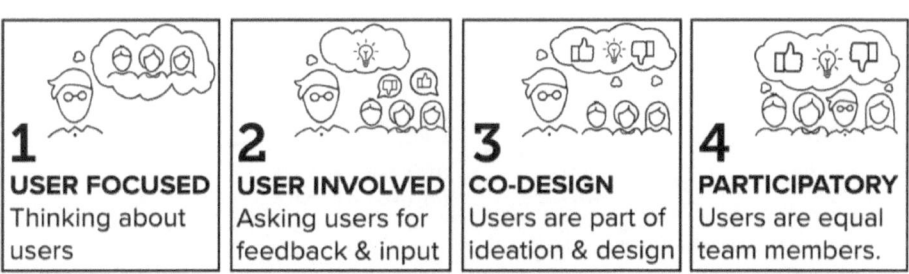

FIGURE 11.1 The range of user contact in user-centred design (UCD) approaches.

levels of empathy and involvement to users participating as team members. Different views on who should receive focus, the degree of direct contact and the user–designer relationship have separated user-centred methodologies into specific approaches, such as inclusive, empathic, co-creative and participatory design. For sensemaking, a mid- to high range of user involvement could probably be fitting, where the continuous involvement of users can provide an in-depth understanding of contextual needs.

Regardless of the chosen degree of direct contact, a situated understanding of user needs and pain points in order to ground the process is needed. Also, at least *some* direct user involvement is expected, particularly in user testing of prototypes, mock-ups and sketches.

UCD AS A PROCESS

The ISO 9241-210:2019 standard presents a generic process model iterating four phases: (1) understand and specify context of use, (2) specify user requirements, (3) produce design solutions to meet user needs and (4) evaluate (validate) designs (ISO, 2019). The ISO model emphasizes iterations but does not address the divergent (exploring) and convergent (selecting) stages that typically take place in a design process – unlike the double diamond model which emphasizes divergence/convergence over visualizing iterations back and forth.

The double diamond is commonly used by design disciplines and emphasizes the process of discovering needs and framing the design problem prior to exploring solutions (Tschimmel, 2012, p. 9). While direct user involvement takes time and effort, relying on indirect user contact runs the risk of creating a product no one needs. The same risk occurs if not enough time is allocated to ensure that the right problem is framed. Using the double diamond in early phases may help scoping and innovating problem definition and framing, including truly understanding the situated context of use. In principle, there is no inherent incompatibility between the double diamond and ISO models, and both are user centred. Their core phases are the same: understand/discover, specify/define, design and evaluate/deliver. The ISO model emphasizes iterations but does not address the divergent (exploring) and convergent (selecting) stages that typically take place in a design process – unlike the double diamond model which emphasizes divergence/convergence over visualizing iterations back and forth.

In UCD efforts, professionals may move between classic stances of positivist, critical and constructivist paradigms. The positivist stereotype seeks information

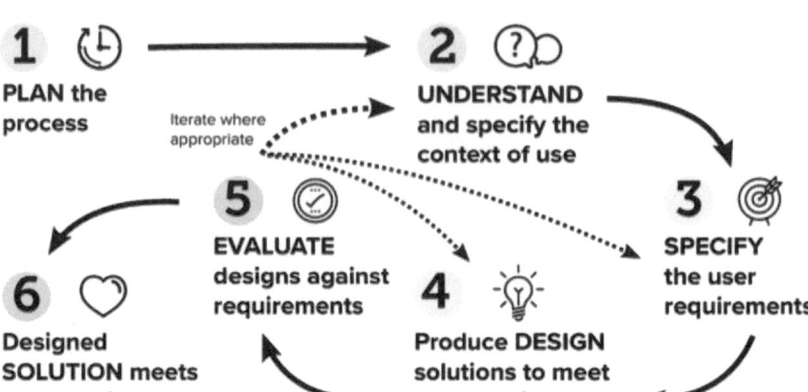

FIGURE 11.2 The user-centred design process.

FIGURE 11.3 The double diamond model.

in order to advice: *What is known? How can we reach our aim?* The critical ste-reotype is advocating for positive change and triggering innovation by questioning: *Is this the right aim? Why?* The constructivist stereotype is a negotiator, aiming for shared insights and efforts: *Who are stakeholders and priorities? How can we collaborate?* Early on, one might feel the need to be more critical and exploratory in order to better frame the problem. There is not a single right way, model or approach to do UCD.

UCD METHODS AND TECHNIQUES

There is a wide and ever-growing range of UCD methods, and it takes experience and reflection to choose the "right" one at each step. The next sections outline some of the most common methods, grouped to correspond to the core UCD phases: (1) understanding, (2) specifying, (3) designing and (4) evaluating. The iterative nature of the UCD process supports the selection of *any method* at *any time*, asking, e.g., What are the most important insights to gain next? How much time and resources do we have for UX work now and later?

Understanding: Discover Insights

Design processes typically start with an exploration of the problem area at hand. A core aspect of UCD is framing the goals in situated user needs. Thus, entering into the context of use to gather rich qualitative insights is recommended. UCD has traditionally been quite task focused on specific usage goals (e.g. Pinelle & Gutwin, 2003), overlapping with UCD. However, UCD also overlaps with anthropology in its field work to discover situated needs and the characteristics of the users. Key questions may be: *Who are our users and their experienced reality, needs and mental models? What is the context and nature of use?* This work is often called "user research" or "design research". You ask, listen and take notes.

Perhaps the most frequently used methods are interviews and observations (Begnum & Thorkildsen, 2015). These methods are flexible and may be contextual, formal or informal, spontaneous or planned, open and involved or closed and hidden. They include both focus groups and task analysis. There are many other methods that can also be utilized, such as workshops, card sort, probes, environmental analysis, empathic modelling and service safaris. Using quantified data from surveys or web analysis is a low contact method for knowledge gathering. Since they seldom help you identify pain points in complex real-life scenarios, they are not recommended as sole methods for finding and framing the right problem to solve.

Specifying: Define Needs

As UCD is iterative, the lines between gathering insights and specifying needs are blurred. You discover something, create assumptions, then go back to investigate interesting findings. Commonly used methods for specifying needs in UCD are affinity maps, personas, value propositions and scenarios-related techniques such as storyboards, user journeys and user stories. Many specifying methods visualizes the elicited insights. Such visual artefacts are recommended to convey user needs to stakeholders and facilitate team communication (Garcia, Silva da Silva, & Silveira, 2017). The visualizations can be updated as new discoveries are made. The specification phase is ended when you have framed your problem and feel certain that you have discovered true contextual needs and pain points.

Designing: Explore Solutions

Moving on to exploring solutions, there is a massive range of techniques available to ideate, create and evaluate design concepts – including blue-skying, brainstorming and bodystorming (D. Gray, Brown, & Macanufo, 2010). Many of the scenario-related techniques cross over to the concept exploration phase. User involvement and co-creation may be utilized to explore alternative design directions early on. They also help uncover user perspectives and additional insights into the underlying emotions and values behind user preferences. Workshops are recommended for interdisciplinary discussions. In addition, role playing and physical prototypes are useful techniques. In order to express design concepts, low-fi prototypes such as sketches, mock-ups, presentations and whiteboarding are commonly used (C. M. Gray, 2016).

Evaluating: Testing and Adjusting

At some point, a selection of one or more design directions is made – for example, through prioritization matrices. Early on, focus should be on validating the value of a concept not on the usability of the design. It is not uncommon to involve users and stakeholders in design walkthroughs (called pluralistic walkthroughs). Increasingly, participatory and co-creative design work is also being used, enabling rapid user feedback and co-discovery tests.

Next, potential solutions are typically tested with users and adjusted in an iterative manner. An iterative process describes a strategy where something is produced, then evaluated and next refined. Low-contact methods are usually utilized as a first step, such as design critique based on design guidelines, heuristic evaluations and viability analyses, before moving on to user testing. Though formal (summative) testing is reportedly used in UCD literature, informal usability tests appear the most common-place in industry. Informal usability testes are conducted as soon as practical – for example, through high-fi prototype guerrilla testing (going to public libraries, sub-way testing, asking cantina personal, taxi drivers, etc. depending on the user envi-ronment). In UCD, usability testing usually refers to user testing. Pilot user feedback is also utilized, as are online testing and online prompting for feedback (analytics tools). Pilot user feedback is also utilized, as are online testing and online prompting for feedback (analytics tools).

A BRIEF INTRODUCTION TO AGILE DEVELOPMENT

The term "agile" is applied to software development processes that are iterative and incremental, focused on continuous improvement of working code while leaving room for exploration and change (Petersen & Wohlin, 2010). An incremental process develops a larger system by partial deliveries. By delivering partial value continu-ously, two things happen: (1) it is possible to get early feedback and find what works and does not work for the user, with less time spent on planning and documenta-tion and (2) through continuous testing and deployment pipelines, costumers and end-users get continuous value output and software quality is constantly improved.

Compared to traditional plan-based and overly requirements-based approaches, the agile mindset is more efficient in quickly and reliably producing a usable system (Beyer, 2010, p. 1). Early partial deliveries are usually not working code (e.g. proto-types). As the codebase matures, changes become smaller, allowing for an efficient automated deployment of the new code known today as continuous delivery or contin-uous deployment. Incremental deployments may include operations-related updates and roll backs, development-related features, bug fixes, redesigns and re-factoring. The agile mindset facilitates a high change tolerance from one iteration to the next, supports rapid exploration of technical solutions, priority adjustments based on feed-back and continuous error fixing (Constantine, 2001).

The agile development aims of efficiency and reduced overhead overlaps with principles of lean development (as presented in, e.g., Hines, Holweg, & Rich, 2004). Some view lean as part of the agile mindset while others argue lean is a separate mindset more focused on production flow. Agile and lean share focus on continuous improvement as well but with different perspectives; agile methods focus on software development while lean is focused on product development in general. In this section, we discuss combining agile and UCD with focus on software development.

AGILE CORE PRINCIPLES

The Agile Manifesto of 2001 declared 12 principles, summarized as: (1) satisfy the customer, (2) welcome changed requirements, (3) deliver software frequently, (4) involve businesspeople daily, (5) trust team members and ensure their motivation, (6) use face-to-face communication, (7) measure progress in working software, (8) keep a sustainable pace, (9) continuous attention to quality, (10) keep it simple, (11) use self-organized teams and (12) do regular effectiveness reflections (Beck et al., 2001). Overarching these were four values: individuals over process, working software over unnecessary documentation, change response over following a plan and customer collaboration over contract negotiations. Since then, the agile values have been updated to the following modern agile principles (Modern Agile):

1. "Make People Awesome"
2. "Make Safety a Prerequisite"
3. "Experiment & Learn Rapidly"
4. "Deliver Value Continuously"

The focus of the Modern Agile community is aiding costumers to be great at what they do, no longer simply on creating a great product (ModernAgile). As such, the separation between the costumer and the team (as a supplier) is minimized over time. Earlier, agile development was ended when the costumer was satisfied with the supplied solution. Today, one recognizes delivery is continuous throughout the lifecycle of a solution. This has inspired DevOps practices of merging development and IT operations by deploying features into production quickly and continuously correct problems. A modern agile process is ended when the decision is made to discard or redevelop a solution.

THE AGILE PROCESS

There are different ways of organizing the agile teamwork to support the agile principles, and as such there are infinite variants of the agile processes. Some popular agile frameworks are Extreme Programming (XP), Scrum and Kanban (Agile Alliance):

• Scrum is a widely known framework. The Scrum model is focused on facilitating self-organizing short push-iterations (sprints) – including efficiency reviews and team roles. The team delivers new software to the costumer at certain time intervals.

- XP is the most specific framework, pushing specific methods and practices such as co-location, pair programming, weekly iterations, continuous integration and quarterly releases.
- Kanban is regarded today as agile but has roots in lean thinking of eliminating bottlenecks and moving from push- to pull-based flows. Kanban does not set time-limited iterations (sprints) and instead focuses on limiting the number of work tasks in progress.

It is common for companies and teams to create their own process models by adjusting the proposed agile frameworks. Overall, iterative feedback loops, rapid learning and incremental software delivery are key agile attributes – as is the role of the team. Agile teams are usually multidisciplinary, and team members organize tasks and work amount in a democratic manner. Though many agile teams apply time-limited iterations that ends with a delivery (Scrum-like), this is not a must. Some agile teams strive to achieve efficient workflows (Kanban-like). Currently, there appears to be a shift from the early agile process approaches towards even lighter frameworks guided by the modern agile principles – without elevating specific practices.

As agile has evolved, focus on risk management and user empathy has also been increased. Modern agile approaches seem less focused on estimating story points for feature development and the efficiency (velocity) of the team to develop features. Instead, the importance of a safe culture for experimenting and learning from mistakes is emphasized. As such, even if efforts are planned with the aim of incremental delivery, there is a shift towards learning of end-user needs through product development iterations. Note that just as in UCD, agile processes are not mainly focused on initial problem framing activities, but rather on experimentation based on assumptions of needs.

USER-CENTRED AGILE DEVELOPMENT (UCAD)

Both UCD and agile mindsets embrace iterative and incremental processes with frequent feedback loops to explore, learn and improve. These practices have proved valuable in delivering the right solutions to the problems at hand. However, there are indications that some user-centred values are overlooked in the agile world. The next sections discuss the three main dangers when merging UCD and agile development: (1) rushing problem framing, (2) interface without context and (3) workload discrepancies and flow issues.

Rushing Problem Framing

User-centred methods are well suited for gathering in-depth insights, identifying needs and scoping the design space based on user needs – including deciding which needs are the most critical to solve and innovating on possible strategies to do so. Though well suited for problem solving, agile settings have less tools for problem framing – including exploring whether software development is the right approach to solve a user need. Compared to the user-centred double diamond model, agile frameworks do not emphasize the early process of problem framing prior to exploring solutions. Experimenting in incremental and iterative development is not the same as problem framing. Singh (2008) notes that agile product goals are sometimes set without ensuring an adequate study of the user's needs and context.

User research and problem framing can be embedded into UCAD. However, many UCAD processes do not emphasize framing or the iterative, divergent and convergent aspects of this phase. UX designers may be expected to quickly gather user insights, identify contextual needs and frame the problem on their own, once and at the start of the process. If the team pushes to start coding, this necessitates the problem at hand is already well defined. If not, the problem at hand may be based on weak or faulty assumptions about users and context of use (Constantine, 2001) – which will also make the solutions to the problem ill-suited for real-life scenarios.

FORGETTING THE CONTEXT

Usage is affected by the social, physical and emotional contexts, by noise, time and space, and by what happens prior to and parallel to any system interactions. These aspects must be taken into consideration both when framing and understanding the problem at hand and in the design and evaluation of the potential solutions. Prototype testing, even if simulating realistic scenarios, may not be sufficient on its own to elicit needs.

It appears UCAD settings have less user contact and utilization of user involvement compared to traditional UCD, emphasize user interface (UI) over other design activities and prioritize feature implementation over in-depth user research (Begnum & Thorkildsen, 2015; Silva da Silva, Silveira & Maurer, 2013). Singh (2008) observes that agile development teams sometimes have difficulties seeing the whole UX and relevant user-centred aspects, and instead focus on just-in-time deliveries. In a worst-case scenario, one may imagine a team holding a faulty understanding of the real issue at hand, a weak understanding of contextual needs and testing solutions in non-realistic settings – such as a laboratory.

Focus on UI without investigating real-life issues may lead to the team being unaware of important contextual aspects influencing the overall UX and usability of a product. This is particularly important with regard to safety-critical solutions used in complex sensemaking settings, where disregarding social, emotional, contextual and timely aspects (e.g. non-verbal communications, expectations, stress, noise, distractions, fatigue) may carry huge risks.

WORKLOAD DISCREPANCIES AND FLOW ISSUES

UCAD also faces challenges related to workload discrepancies, interdisciplinary collaboration and flow issues. At least five generic models have been suggested for UCAD, aiming to mitigate some of the issues described in the previous sections. The models are not necessarily mutually exclusive nor are they exhaustive on possible ways to organize UCAD workflows.

Parallel Model

In the *parallel model*, UX work and coding is separated, forming two different "tracks". It has also been referred to as the "dual track agile" model and proposes doing UX work ahead of development (Miller, 2005; Sy, 2007).

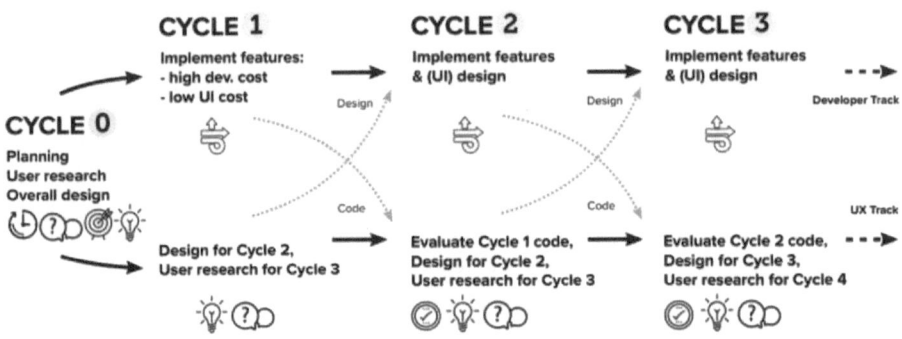

FIGURE 11.4 The parallel model.

The first thing to note is that though the model starts with a Cycle 0, encompassing user research, problem exploration and framing are not emphasized in the parallel model. Cycle 0 focuses on planning how to build an already agreed-upon problem and includes ensuring sufficient insights into contextual user needs is held by the team. If a UX designer starts questioning the planned solutions, this is an indication you have rushed the problem framing.

The second thing to note is that the parallel model does not remind the team to remember contexts of use. In fact, the UX track was originally labelled the "Interaction Designer track" by Sy (2007) and focused on UI design. With a UI focus, Cycle 0 is likely to be used as an iteration for planning the overall UI design of an agreed-upon concept.

The third thing to note is that the team must plan several sprints ahead in this model and as such is less adaptable to change. Any changes could necessitate rework for the UX track. The UX track is at last one sprint ahead of the coding track, conducting contextual inquiries, prototyping and user testing prior to passing on validated designs ready for implementation to the development track. If the coded designs need further user testing, this would also be UX work, and as such a designer would work both ahead of and behind the developers. This has been described as exhausting by designers (anecdotal evidence, unpublished interviews).

Finally, the separation of UX and development split the team – with developers often leaving all the UX work to one designer. Thus, two parallel processes go on at the same time, with iterating deliveries from one track to the other. The interaction between these two tracks is structured and frequent, but if separated into developers waiting for UX input, there is a risk of "UX bottlenecks" hindering workflow due to workload discrepancy. Also, parallel work organization hinders interdisciplinary discussions and problem solving. Parallel models thus appear the most advantageous when there are adequate resources to secure collaborative team efforts within a sustainably paced work environment, with the team pulling together on both tracks and where there is UCD competence (e.g. with a designer guiding team UX work).

Satellite Model

The idea of the *Satellite model* is to solve the issue of a single UX resource per development team becoming exhausted or a bottleneck. Here, one satellite UX person is co-located with an agile team as a UX specialist, focused on supporting the agile development. The UX specialist also facilitates collaboration with the full UX team (Kollmann, Sharp, & Blandford, 2009). The UX team does ideation, user involvement, prototyping and usability testing on several projects, as a joint effort. The UX team may pull backlog items directly, disconnected from development, and may work on different features. Apart from the UX specialist, the UX team is separate from the development team in this model (Øvad, 2014). One worry is that incorrect assumptions may be made by the UX team or the developer team. As such, the model improves work balance and flow between development and UX – but does not create opportunities for interdisciplinary collaboration.

The UScrum Model

Singh (2008) proposed the *UScrum model* where an UX management role is added to Scrum, in order to solve the challenges of problem framing and understanding contextual needs as well as ensuring user-centred tasks for ensuring that usability is not de-prioritized. To do this, UScrum splits the Scrum product owner role in two where one takes on an explicit usability focus and the other focuses on feature implementation. With regard to modern agile thinking, the idea that a product owner is feature-focused feels outdated. A product owner is expected to provide the team with insights into business *and* user needs and help with work prioritizing. Furthermore, the traditional agile focus on making the customer *happy* is changed to *awesome* in modern agile – increasing end-user focus. Nonetheless, UScrum represents a way to improve work balance and flow between development and user-centred perspectives without splitting into two teams or tracks, thus supporting interdisciplinary collaboration.

Lean UX Model

A newer model for integrating UCD and agile mindsets is *lean UX*. This model relies on modern agile values and aims to support moving to iterations based on *assumptions* instead of creating an established to-do backlog with user stories (Øvad, 2014).

FIGURE 11.5   The satellite model.

UScrum model, based on Singh (2008)

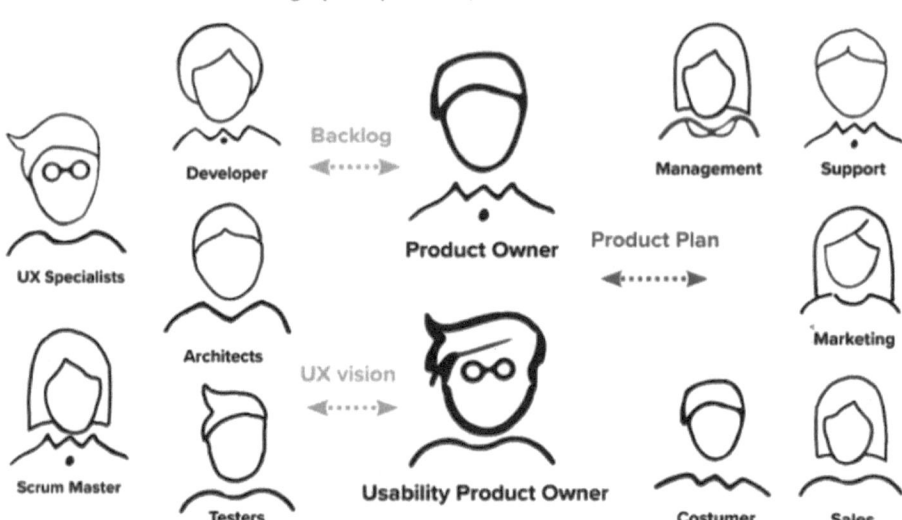

FIGURE 11.6   The UScrum model.

Lean UX proposes four steps: (1) research, (2) declare assumptions, (3) create a solution (based on the assumptions), and (4) test (measure) the solution (and assumptions) through an experiment. The shift from stating a requirement to declaring an assumption clarifies the need for contextual experimentation. The model thus aligns well with the UCD model and communicates better how decision points are constantly needed for clarification on the viability of a solution. The model emphasizes the mindset of experimenting and learning, rather than the incremental delivery, and facilitates interdisciplinary collaboration.

User-Centred Agile Model

The user-centred agile model combines an initial user-centred process to help understand the user and context, discover needs and frame the problem with agile development to iteratively and incrementally experiment, build, learn, deploy and improve software. The user-centred agile model uses the double diamond model for initial UCD and the lean UX model for agile development, and as such combines UCD, design thinking, lean and agile approaches.

This model encompasses both initial and continuous user-centred explorations, focusing both on building the right thing and building the thing right. Compared to the parallel model, user research and user involvement is now explicitly integrated as part of the UCAD process. This combats the reduction of UCD to UI design and the subsequent omittance of physical, social and emotional contexts of use and sensemaking. Furthermore, the user-centred agile model prevents problem framing from being skipped. Implicit in the model is the belief that the team only moves into lean UX if they believe that software development can solve the problem at hand.

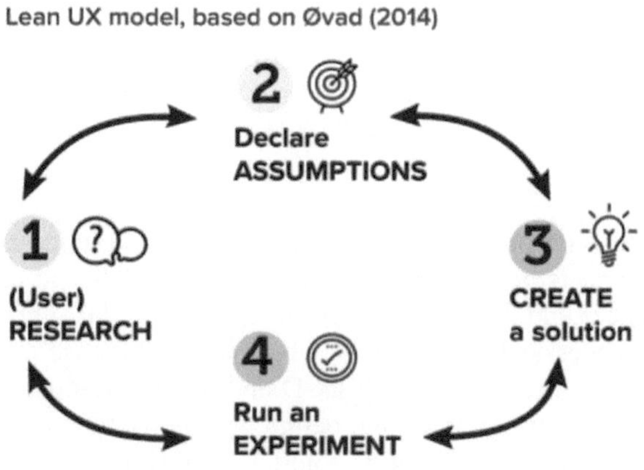

Lean UX model, based on Øvad (2014)

2 Declare
ASSUMPTIONS

1 (User)
RESEARCH

3 CREATE
a solution

4 Run an
EXPERIMENT

FIGURE 11.7 The lean user experience (UX) model.

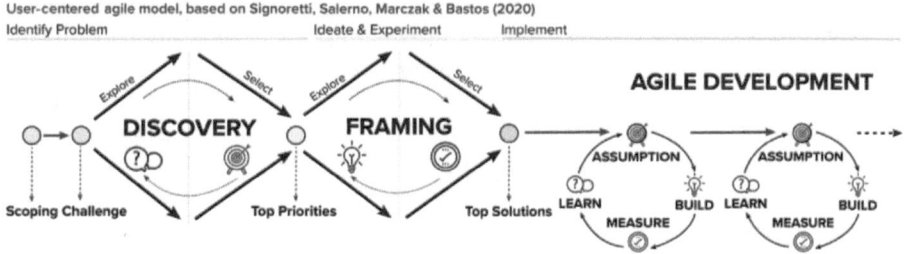

User-centered agile model, based on Signoretti, Salerno, Marczak & Bastos (2020)

Identify Problem   Ideate & Experiment   Implement

FIGURE 11.8 The user-centred agile model.

A recent case study documented the effect of implementing the user-centred agile model and showed that by involving the whole team in framing stages (such as user interviews, user research and problem exploration), there was a shift from focusing on requirements to map user and business problems (Signoretti, Salerno, Marczak, & Bastos, 2020). This promoted a problem-oriented mindset and shared ownership. This mindset merged well with the assumption-based lean UX iterations, as teams reported they felt safer to fail.

Another finding in their work was that the overall user focus was strengthened. The individual team members reported feeling responsible for investigating the problem at hand and actively seeking to understand user needs. Further, this shared UX responsibility resulted in increased cross-disciplinary decision-making processes, and it particularly influenced the role and collaborative attitude of the developers. Overall, teams reported increased engagement, user empathy and a shared project vision (Signoretti, Salerno, Marczak, & Bastos, 2020).

As such, preliminary research indicates the user-centred agile model is successful in supporting UCAD teams in strengthening problem-oriented mindsets, collaborative decision-making and user empathy. Both from theoretical assumptions and empirical evidence the model shows promise in relation to alleviating the three main UCAD challenges reported by research: (1) rushed problem framing, (2) forgetting the context and (3) workload discrepancies and flow issues.

FINAL REFLECTIONS

Neither UCD nor agile presents a one-size-fits-all best-practice approach. Instead, one draws on mindsets, methodologies and process models and adapts these to the problem, the context and the available resources at hand. An approach fitting in one setting does not necessarily fit the next. The same is the case for UCAD. As long as you adhere to the mindsets of both UCD and agile development, you can safely draw on a wide range of practices and define workflows that suit your needs while remaining certain that you are creating solutions that work for your users in real life and real context of use.

Literature reports a culture gap between software designers and developers, where diverging core values create challenges in collaboration and communication (e.g. Bhrel, Meth, Maedche & Werder, 2015; Beyer, Holtzblatt & Baker, 2004; Salah, Paige & Cairns, 2014). Bhrel, Meth, Maedche and Werder (2015) present an extensive literature survey on UCAD including 84 research papers in order to define a common set of UCAD principles. They are nonetheless unable to derive any principles related to team organization, collaboration, communication, decision-making, roles and responsibilities. Their conclusion is that literature does not present a clear alternative to teamwork organization as parallel tracks, even though there is very limited empirical evidence supporting the one sprint ahead concept. Moreover, they note that research does not point out two separate teams in parallel as being the optimal. About 41% of the included papers described cross-disciplinary agile teams – and not two dedicated teams handling UX and development. They paraphrase Kuusined (2012) who reports that the most frequent advice from practitioners on improving cooperation, decision-making and knowledge transfers is having UX competence available within the team (2014). Salah, Paige and Cairns (2014) surveyed 71 papers and recommend (1) sharing an understanding of users, (2) sharing an understanding of design vision, and (3) synchronization of parallel efforts to allow for collaboration (e.g. daily synchronization points) in order to optimize workflows and cross-disciplinary relationships.

On the topic of building the right product vs. building the product right, Bhrel, Meth, Maedche and Werder (2015) conclude that research supports a principle of separating product discovery and product creation. They reference Kettunen (2009) who proposed drawing on UCD to counter agile shortcomings with regard to large-scale product innovations, scoping or ideation. However, they do not specify whether the agile team should be involved in both product discovery and product creation or whether they envision a handoff from discovery to the agile team. Beyer, Holtzblatt and Baker (2004) recommend that agile teams use rapid contextual design (rapid CD) to support problem framing and design and present a stepwise approach with specified methods that appear detailed, but otherwise align with UCD and the double diamond model. However, they reduce UX work to UI design and envision the skillsets of design and development remain separated – similarly organized as in the parallel model. They propose costumer(s) and designer(s) complete agile rapid CD, subsequently handing off completed UIs and user stories to development. From the extensiveness of the rapid CD process, workflow bottlenecks are a risk and developers would not be involved in decision-making and design choices based on their process descriptions.

It is interesting that any disciplinary culture gap seems to be bridged by applying the user-centred agile model, where Signoretti, Salerno, Marczak and Bastos (2020) documented that both user focus and problem orientation were increased and that the UX responsibility and decision-making was shared. We now have an alternative to the parallel model for teamwork organization that fits better with recommendations from Salah, Paige and Cairns (2014). It is also worth noting that the gap identified in research appears to decrease over time, as modern agile and UCD principles and practices become better aligned.

Efforts are ongoing on how to integrate UCD into agile settings; however, these quickly appear dated. Alternative visualizations related to the user-centred agile model can be found on online blogs from around 2018, where design thinking, lean and agile frameworks are combined. However, most of them are focused on merging lean UX with agile and not including UCD specifically. Though the blog posts hold very interesting reflections, they so far appear only anecdotal from the writer's point of view and without scientific publications demonstrating their promise.

Based on the evidence at hand, we would generally caution against parallel tracks and advise to move towards more integrated teams where both UX work and code work are shared by the whole team, in order to ensure better collaboration and flow. Parallel workflows, proposed by previous research on integration models, may be counterproductive to efficient multidisciplinary team collaboration due to the separation of UX/UI design and development decisions. Furthermore, the risk for workload discrepancies may lead to ineffectiveness, communication issues and undermined design decisions. A parallel workflow setup needs generous constraints to ensure a smooth flow and cooperation between iterative UI design, development and UX work.

User-centred agile Development
Dangers & 👍Advice

1 Danger: Rushed problem framing

Advice: **User research & problem framing is team effort**

2 Danger: Omitting the context of use

Advice: **Contextual (in situ) user testing and validation**

3 Danger: Workload discrepancies & flow issues

Advice: **Interdisciplinary decisions & team collaboration**

FIGURE 11.9   Summary: current dangers and advice.

Instead, a closer and continuous cross-disciplinary collaboration is suggested, focusing on a productive team-driven workflow and interdisciplinary decision-making. Problem framing and user research are recommended as *team* efforts. User testing and contextual (in-situ) design validation are recommended as *team* activities. A negative consequence of downplaying UX design is that the *team* gets a weaker understanding of the problem to be solved and the real user needs in real-life scenarios – which is especially dangerous with regard to safety critical systems and complex sensemaking settings.

REFERENCES

Agile Alliance. *Agile Essentials: Agile 101*. https://www.agilealliance.org/agile101/

Beck, K., Beedle, M., Bennekum, A. V., Cockburn, A., Cunningham, W., Fowler, M., ... Thomas, D. (2001). *Manifesto for Agile Software Development*. http://AgileManifesto. org (accessed 21.02.2021); https://moodle2019-20.ua.es/moodle/pluginfile.php/2213/ mod_resource/content/2/agile-manifesto.pdf (accessed 21.02.2021).

Begnum, M. E. N., & Thorkildsen, T. (2015). Comparing User-Centred Practices in Agile Versus Non-Agile Development. *Paper presented at the Norsk konferanse for organisasjoners bruk av IT (NOKOBIT)*, Ålesund, Norway.

Beyer, H. (2010). *User-Centered Agile Methods*. Morgan & Claypool.

Beyer, H., Holtzblatt, H., & Baker, L. (2004). An Agile Customer-Centered Method: Rapid Contextual Design. In Zannier E. H., Lindstrom L. (Ed.), *Extreme Programming and Agile Methods - XP/Agile Universe 2004*, LNCS vol. 3134, 50–59. Springer.

Bhrel, M., Meth, H., Maedche, A., & Werder, K. (2015). Exploring principles of user-centered agile software development. *Information and Software Technology, 61*(C), 163–181.

Constantine, L. L. (2001). Process agility and software usability – towards lightweight usage centred design. *Information Age, 8*(8), 1–10.

Forlizzi, J., & Battarbee, K. (2004, August 1–4, 2004). Understanding Experience in Interactive Systems. *Paper presented at the DIS2004*, Cambridge, MA.

Garcia, A., Silva da Silva, T., & Silveira, M. S. (2017). Artifacts for Agile User-Centered Design: A Systematic Mapping. *Proceedings of the 50th Hawaii International Conference on System Sciences (HICSS)*.

Gould, J. D., & Lewis, C. (1985). Designing for usability: key principles and what designers think. *Communications of the ACM, 28*(3), 300–311.

Gray, C. M. (2016, May 7–12). It's more of a Mindset Than a Method: UX Practitioners' Conception of Design Methods. *Proceedings of the 2016 CHI Conference on Human Factors in Computing Systems*, San Jose, CA.

Gray, D., Brown, S., & Macanufo, J. (2010). *Gamestorming: A Playbook for Innovators, Rulebreakers, and Changemakers*. Sebastopol, CA: O'Reilly.

Hartson, R., & Pyla, P. (2012). *The UX Book; Process and Guidelines for Ensuring a Quality User Experience*. Waltham, MA: Elsevier.

Hines, P., Holweg, M., & Rich, N. (2004). Learning to evolve: a review of contemporary lean thinking. *International Journal of Operations & Production Management, 24*(10), 994–1011. doi:10.1108/01443570410558049

ISO. (2019). ISO 9241-210:2019 Ergonomics of human-system interaction. In *Part 210: Human-centred design for interactive systems*. www.iso.org

Kollmann, J., Sharp, H., & Blandford, A. (2009). The Importance of Identity and Vision to User Experience Designers on Agile Projects. *2009 Agile Conference*.

Miller, L. (2005). Case Study of Costumer Input for a Successful Product. *Paper presented at the Proceedings of Agile Conference (AGILE)*, Denver, CO.

Modern Agile. http://modernagile.org/.

Øvad, T. (2014). *The Current State of Agile UX in the Danish Industry: The Analysis.* p. 33. https://vbn.aau.dk/ws/portalfiles/portal/209798499/The_Current_State_of_Agile_UX_ in_the_Danish_Industry.pdf (accessed 21.02.2021).

Petersen, K., & Wohlin, C. (2010). The effect of moving from a plan-driven to an incremental software development approach with agile practices, An industrial case study. *Empirical Software Engineering, 15*(6), 654–693. doi:10.1007/s10664-010-9136-6

Pinelle, D., & Gutwin, C. (2003). Task analysis for groupware usability evaluation: modeling shared-workspace tasks with the mechanics of collaboration. *ACM Transactions on Computer-Human Interaction, 10*(4), 281–311.

Pruitt, J., & Grudin, J. (2003). *Personas: Practice and Theory.* DUX '03: Proceedings of the 2003 conference on Designing for user experiences, in San Francisco California June 2003, pp. 1–15, https://doi.org/10.1145/997078.997089.

Rubin, J., & Chisnell, D. (2008). *Handbook of Usability Testing, Second edition: How to Plan, Design, and Conduct Effective Tests* (2nd ed.). Indianapolis, IN: Wily.

Salah, D., Paige, R., & Cairns, P. (2014). A Systematic Literature Review for Agile Development Processes and User Centred Design Integration. *18th International Conference on Evaluation and Assessment in Software Engineering (EASE '14)*, London, England.

Sengers, P., Boehner, K., David, S., & Kaye, J. J. (2005, August 21–25, 2005). Reflective Design. *CC '05-4th Decennial Conference on Critical Computing: Between Sense and Sensibility.* Århus, Denmark.

Signoretti, I., Salerno, L., Marczak, S., & Bastos, R. (2020). Combining user-centered design and lean startup with agile software development: a case study of two agile teams. In: Stray V., Hoda R., Paasivaara M., Kruchten P. (eds) *Agile Processes in Software Engineering and Extreme Programming. XP 2020.* Lecture Notes in Business Information Processing, vol 383. Cham: Springer.

Silva da Silva, T., Silveira, M. S., & Maurer, F. (2013). Ten Lessons Learned from Integrating Interaction Design and Agile Development. *Agile Conference (AGILE)*, Nashville, TN.

Singh, M. (2008). U-SCRUM: An Agile Methodology for Promoting Usability. *Agile 2008 Conference.*

Sy, D. (2007). Adapting usability investigations for agile user-centered design. *Journal of Usability Studies, 2*(3), 112–132.

Tschimmel, K. (2012). Design Thinking as an Effective Toolkit for Innovation. *The International Society for Professional Innovation Management (ISPIM)*, Manchester.

Vetere, F., Gibbs, M. R., Kjeldskov, J., Howard, S., Mueller, F. F., Pedell, S., … Bunyan, M. (2005, April 2–7, 2005). Mediating Intimacy: Designing Technologies to Support Strong-Tie Relationships. *Paper presented at the CHI 2005*, Portland, OR.

12 Improving Safety by Learning from Automation in Transport Systems with a Focus on Sensemaking and Meaningful Human Control

Å. S. Hoem
Norwegian University of Science and Technology

S. O. Johnsen
SINTEF

K. Fjørtoft and Ø. J. Rødseth
SINTEF Ocean

G. Jenssen and T. Moen
SINTEF

CONTENTS

INTRODUCTION

There is an increase in the use of automation and autonomous solutions within transportation. According to *The Oxford Dictionaries*, autonomy is the right or condition of self-government, and the freedom from external control or influence. Many researchers (Relling et al., 2018) have discussed that the term is used differently in colloquial language than in the technical definition and that it is interpreted in different ways across industries. In this chapter, we emphasise that autonomy does not necessarily mean absence of human interaction. Often there is a strong need to design how humans can make sense of automation failures and enact meaningful human control.

Automated systems operate by clear repeatable rules based on unambiguous sensed data. An autonomous system can be a set of automated tasks, with interactions with several sub-systems and/or humans, with a specific degree/level of autonomy. Autonomous systems obtain data about the unstructured world around them, process the data to generate information and generate alternatives and make decisions in the face of uncertainty. Systems are not necessarily either fully automated or fully autonomous but often fall somewhere in between (Cummings, 2019). For example, transportation can have different modes during a sea voyage. Outside the harbour, in heavy traffic, it can be closely operated either by the remote control centre (RCC) or a captain/driver, while in open waters with low traffic it can be controlled by the computers or the autonomous system. Within the road traffic segment, the Society of Automotive Engineers (SAE) has defined a taxonomy on the levels of automation describing the expectations between automated systems and the human operator (SAE, 2018). This is summarised in Table 12.1 below.

The levels apply to the driving automation feature(s) that are engaged in any given instance of operation of an equipped vehicle. As such, a vehicle may be equipped with a driving automation system that is capable of delivering multiple driving automation features that perform at different levels. The level of driving automation exhibited in any given instance is determined by the feature(s) that are engaged (SAE, 2018). Hence, autonomy is different across application areas; it varies over time and is affected by the context.

To get a better overview and understanding, we start by looking at experiences gained from ongoing research and/or industry projects in the four transportation

TABLE 12.1

Levels of Automation – Simplified Description from SAE J3016 (2018).

LoA	Humans in control	Automation in control	Examples of automated features
0: No driving automation	All operations	No automated task. Warns; protect	Blind-spot monitoring and lane-departure warning
1: Driver assistance	All operations	Single automated systems: assists	Adaptive cruise control (ACC)
2: Limited assist; auto throttle	Drives in-the-loop	Guides	Automated lane centring combined with ACC
3: Assist, tactical; supervised	On-the loop human monitors all time	Manage movement within defined limits	"Traffic jam chauffeur"
4: Automated assist strategic	Out-of-loop asked by system	Operates, but may give back control	Self-driving mode with geofencing
5: Autonomous	Completely out-of-loop	Operates with graceful degradation	None are yet available to the general public

domains: road, sea, rail and air. Through these case studies, we aim to explore safety, security, sensemaking and the human control of autonomous transport systems. We have adopted the term "meaningful human control" from discussion and debates from another area (lethal autonomous weapon systems; Cummings, 2019). The term addresses the concerns of a "responsibility gap" for harms caused by these systems, i.e. humans, not computers and their algorithms should ultimately remain in control of, and thus morally responsible for, relevant decisions about military operations. The same concern must be the result of autonomous systems in transportation, i.e. humans (supported by computers and algorithms) should ultimately remain in control and responsible for relevant decisions. The responsibility may be on the designer and producer of the autonomous systems, as Volvo and Mercedes Benz have stated for their autonomous cars (Chinen, 2019, p. 109).

BACKGROUND: SAFETY OF AUTONOMOUS SYSTEMS

Safety is commonly defined as freedom from unacceptable risk (Hollnagel et al., 2008). For autonomous transportation to become a success, It must prove to be at least as safe and reliable as today's transport systems. By some, it is claimed that increased safety will be achieved by reducing the likelihood of human error when introducing more autonomy (Ramos et al., 2018). However, autonomy may create new types of accidents that before were averted by the human in control, as demonstrated by the Tesla fatal accident with Joshua Brown, NTSB (2017). Besides, the introduction of new technology will create new accident types, as explained by Porathe et al. (2018), Teoh and Kidd (2017), and Endsley (2019). The main safety challenges for autonomous systems are unexpected incidents not foreseen by automation, cybersecurity threats, technological changes (with increased complexity and couplings), poor sensemaking, lower possibility for meaningful human control (Human not in the loop) and limited learning from accidents.

The term "Human in the loop" means that the human is a part of the control loop, i.e. that the human receives information and can influence other parts of the chain of events (Horowitz and Scharre, 2015). A key issue is the ability of the actors to make sense of the situation. In our study, we define sensemaking in a pragmatic context as a continuous process of interpreting cues to establish situational awareness in a social context, as described in Kilskar et al. (2020).

When trying to scope risks of autonomous systems, we must include regulation, risk governance, organisational framework, interfaces between humans and the autonomous system, and the available infrastructure (software components and cyber-physical systems) to build a sense of the situation for humans and the automated system (Johnsen et al., 2019).

Autonomous systems are socio-technological systems. Hence, a holistic approach is necessary, rather than a reductionist approach looking at the system as isolated processes and components. We lack statistical evidence for the probability of accidents with autonomous transportation systems. However, several actors have started pilots with different levels of autonomy within different transport modes. There is a need to collect and systemise experiences from these. The following sections present a review of experiences from different transport modes. The main objective has been to gather experiences and status on different transport domains and to learn between the modes, by asking the following research questions:

1. What are the major safety and security challenges of autonomous industrial transport systems?
2. What can the various transport modes learn from each other regarding safety and security related to sensemaking and meaningful human control?
3. What are the suggested key measures related to organisational, technical and human issues?

FINDINGS

AUTONOMY AT SEA

Several countries have developed test areas for testing Maritime Autonomous Surface Ships (MASS). The International Maritime Organisation (IMO) currently uses the term MASS for any vessel that falls under provisions of IMO instruments and which exhibits a level of automation that is currently not recognised under existing instruments. There are already several small-size unmanned and autonomous maritime crafts which have been engaged in surface navigation, scientific activities, underwater operations and specific military activities.

In Norway, three national testing areas have been established, with supporting infrastructure, with the aim to test out MASS in the same area as conventional ships. Norwegian Forum for Autonomous Ships (NFAS, 2020) is a network established for sharing experiences and research within the subject of autonomous ships, with the International Network for Autonomous Ships (INAS, 2020) as an extension of NFAS outside Norway. The research centre for Autonomous Marine Operations and Systems (AMOS, 2020) at NTNU was established in 2013 as a multidisciplinary centre for autonomous marine operations and control systems.

More extensive research projects, such as AAWA (2020), MUNIN (2020), Autosea (2020), Autoship (2020) and IMAT (2020), focus on specific concepts where unmanned, autonomous or smart ships are explored and tested. The world's first fully electric and autonomous container ship, Yara Birkeland (2020), is under construction. The ship is now planned to be in operation by 2022, earlier planned to start in 2020, and centres are scheduled to handle all aspects of remote and autonomous operation to ensure safety.

A newly established company, Zeabuz (2020), will test prototypes of an autonomous electric ferry system for urban waterways. Limited information is given about the concept other than it will be self-driving and electric. The remote and autonomous operational aspect of an RCC is not mentioned, but a remote support center is planned to operate in the initial phase.

Most of the projects above are in the initial stages with limited operational experience. Most safety concerns are related to the reliability of sensors and technical equipment and their ability to handle different situations.

Experiences Related to Safety Challenges

In operation, MASS have only been tested in small scale without an interface for human supervision or control. We have examples of safety issues during early testing of autonomous technology (software and hardware) local in Norway in Trondheimsfjorden, with the small-scale version of the passenger ferry *AutoFerry*. One example is loss of control due to a technical failure, a so-called fallout, of the dynamic positioning system which made *AutoFerry* run into the harbour. However, there is no systematic data collection of failures or unforeseen events, and this is not a requirement from the Norwegian Maritime Authority (NMA) at present. Though, a Preliminary Hazard Analysis (PHA) has been carried out for the operation of the *AutoFerry* (Thieme et al., 2019), the main hazards were software failure; failure of internal and external communication systems; traffic in the channel (especially kayaks, difficult to discover); passenger handling and monitoring; and weather conditions. The practical challenges encountered in the ferry project were also listed. These challenges are related to available risk analysis methods and data, determining and establishing an equivalent safety level, and some of the prescriptive regulations currently in use by NMA. At present (start 2021) the *AutoFerry* project lacks an established plan on who should operate the ferry and how to intervene especially during emergencies. The human operator is said to be in the loop and able to intervene from an RCC. However, none of the projects have developed such a centre or made detailed plans for their operation so far. In the reviewed projects, the focus has been on technology development.

A literature review on risk identification methods for MASS (Hoem, 2019) identifies the uncertainty of the operational mode and context of the MASS operation (i.e. operational domain) to be a major challenge when identifying operational hazards and risks. There is a need to define what conditions the ship is designed to operate under. Rødseth (2018) proposed to use the "operational design domain" from SAE J3016 (2018) to define the context, i.e. the operational domain with its complexity.

This term is further described as an operational envelope (Fjørtoft and Rødseth, 2020). An operational envelope defines precisely what situation the MASS must be able to handle by assigning responsibilities to the human operators and the automation. It defines conditions of operations, describes the characteristics and requirements of the system and enables the design of Human–Autonomy Interface (HAI), based on specific task analysis, safety-critical tasks and challenges of sensemaking.

Several different guidelines are developed for autonomous shipping. IMO has published an Interim Guideline for MASS trials which aims to assist authorities and relevant stakeholders to perform autonomous tests. It includes risk management, how to comply with existing rules and regulations, safe manning, the human element and HMI, infrastructure, trial awareness, and communication and information sharing.

Lessons Learned from Autonomy at Sea

Based on the preliminary testing and risk analysis, it is evident that MASS is a system of systems, depending on local sensor systems, automated port services, communication with RCC, other autonomous ships, conventional ships, Vessel Traffic Centres (VTS) and similar. These interactions are critical factors and should be addressed in design and operations. The degree of autonomy varies and is affected by the complexity of the operation. A MASS will operate in phases with transitions between human control and automation control. A well-defined operational envelope is key for addressing safety issues and carrying out a risk assessment. Potential hazards within each transition must be identified with fallback procedures in place, with focus on the sensemaking process and how humans should enter the control loop.

Challenges related to communicating the intent of a MASS in interactions between autonomous, unmanned ships and manned ships are addressed by Porathe (2019). The authors argue for "automation transparency" and methods allowing other seafarers to "look into the mind" of the autonomous ship, to see if they themselves are detected, and the present intentions of the MASS, i.e. sensemaking among all actors. This can be done by sharing information about the intention, what the automation knows about its surroundings, what other vessels are observed by its sensors and similar by a live chart screen accessible on-line through a web portal by other vessels, VTS, coastguard, etc. Such a common system could be the responsibility of the VTS and should be specified as a requirement for the operational design domain and the operational envelope.

In a guideline from the Bureau Veritas (2019), several hazards are listed as important: voyage, navigation, object detection, communication, ship integrity, machinery and related to systems, cargo and passenger management, remote control and security. Within each of them, a list of factors is mentioned. Using this, Hoem et al. (2019) identified a list of hazards comparing autonomous and manned ships. The scenarios were focussed on the following differentiating factors: fully unmanned, constrained autonomy, RCC, higher technical resilience and improved voyage planning. The paper gave a draft attempt to classify risk factors that can either be characterised as new types of incidents caused by technology, what is most characterised in regard to today's incidents in shipping and if the incidents are averted by crew today. As an example, the category fully unmanned points to a higher risk for technical failure but may improve some of today's operators' errors caused by poor design and lack of good human factor engineering practice. Important factors moving forward are robust sensor quality, redundancy on key technology and good education for land-based operators that support sensemaking and build situational awareness. It is likely that humans are not continuously monitoring one vessel at a time but will be needed to supervise and intervene when necessary. For a constrained autonomous vessel, the paper pointed to the need for better HAI due to the need of time to support sensemaking and get situational awareness before action.

AUTONOMY IN AIR

Automation and autonomy in aviation have been implemented since World War II, where functions have been systematically automated and the manning has been systematically reduced. Incidents due to automation happen, but aviation safety (commercial passenger traffic) is extremely high.

In addition to increased automation in manned flights, the use of drones or unmanned aerial systems (UAS) has risen significantly in the last years. Examples of use are:

- Photography and video recording to support information and crisis management
- Inspection of (critical) components to improve safety, avoid human exposure, reduce costs or improve quality
- Detection and survey of environmental issues, such as gas emissions, ice detection in sea, overview and control of pollution
- Logistics – delivery of critical components or supplies (such as medicine, blood)

Safety Challenges

Manned flights have a high level of safety, issues have often been a result of poor sensemaking and poor situational awareness of the crew. The reliability of the technical equipment is high. Automation accidents have happened lately where guidelines during design and certification have not been followed. This was the case in the Boeing 737 MAX fatal crashes (Cruz and de Oliveira Dias, 2020). After analysing the accidents, Endsley (2019) recommended ensuring compliance with human factors design standards and support for human factors assessment in aircraft testing and certification.

Safety challenges in UAS differ from the challenges in manned operations, due to the immaturity of technology. Looking at the use of large drones in the US, Waraich et al. (2013) documents that mishaps may happen more frequently (i.e. 50–100 mishaps occur every 100,000 flight hours vs human-operated aircraft where there is one mishap per 100,000 flight hours). The mishap rate is 100 times higher in UAS remotely piloted than in manned operations. The leading causes are poor attention to human factors science, such as poor design of human machine interfaces in ground control centres (Waraich et al., 2013; Hobbes et al., 2014).

In Petritoli et al. (2017), the mean time between failures (MTBF) estimated for UAS was around 1,000 hours, approximately 100 times higher than MTBF in manned flights. The dominant failures were in power systems, ground control system and navigation systems.

The risks of UAS operations are dependent on the operational domain, i.e. the type of operation (delivery, data collection, surveillance, inspection photography, etc.) and physical details of the drone such as weight, speed and height of operation. EASA (2016) has estimated the probability of fatality of different UAS weights and estimated probability of fatality as 1% with a UAS weight of 250 g, but 50% fatality with a weight of 600 g in case of a collision with a human when the drone drops.

Examples of undesired incidents from UAS are: collisions with personnel; interference with infrastructure (infrastructure such as airports is vulnerable and interference may lead to disruption of air traffic); actual damage to critical infastructure; damage to the drone; using the drone to spy or steal data (leading to loss of privacy, data theft and possible emotional consequences). Automated systems and UAS are vulnerable to attacks through the cyber-physical systems it consists of, such as sensors, actuators, communication links and ground control systems. As an example, an Iranian cyber warfare unit was able to land a US UAS based on a spoofing attack modifying the GPS data (Altawy et al., 2017).

There are several challenges of UAS operations in challenging climatic conditions such as low temperature, wind, winter with sleet and snow. Operational equipment may not be tested or hardened for these demanding conditions; thus, requirements, testing and certification are needed. Communication infrastructure is also demanding in the north, from 70° the quality of satellite communication is degraded. GPS spoofing may be a challenge and must be mitigated.

Lessons Learned That May Be Transferred

Automation in aviation has succeeded in establishing a high level of safety, due to systematically automating simple tasks and reducing demands on the pilot: base development on the science of human factors, building infrastructure, to control and support flights, strong focus on learning from small incidents and accidents and support from control centres that have strict control of the operational domain/operational envelope. Thus, systematic development and stepwise refinement has had a huge success in terms of safety and trust, in addition to the strong focus on keeping the human in the loop supported by sensemaking. Even in this environment of high reliability, there is a strong need to ensure compliance with human factors design standards and support for human factors assessment in aircraft testing and certification to avoid fatalities by automation as seen in the Boeing 737 Max accidents.

The reliability of drones is lower than for manned planes, and there is a need to develop improved reliability of the new technology. Systematic risk assessment is needed to mitigate the areas with the most risks. The HMI between automation and the human operator is challenging. Design must use best human factors practices to support sensemaking and ensure that the operator can intervene and take control when needed.

Autonomy in Rail

By automated metros (rail systems), we mean systems where there is no driver in the front cabin, nor accompanying staff, also called Unattended Train Operation

(UTO). UTO has been in operations since 1980. According to UITP (2013), there is 674 km of automated metros consisting of 48 lines in 32 cities. Examples of cities with UTOs are Barcelona, Copenhagen, Dubai, Kobe, Lille, Nuremberg, Paris, Singapore, Taipei, Tokyo, Toulouse and Vancouver. There is large infrastructure cost to ensure safe on and offloading of passengers and that the track is isolated from other traffic. Four distinct levels of automation are defined:

GoA1: Non-automated train operation, with a driver in the cabin.

GoA2: Automatic train operation system controls train movements, but a driver in the cabin observes and stops the train in case of a hazardous situation.

GoA3: No driver in the cabin but an operation staff on board.

GoA4: Unattended train operation, with no operation staff on board.

Safety Challenges

Wang et al. (2016) list the following as arguments for UTO: increased reliability, lower operation costs, increased capacity, energy efficiency and an impressive safety record. We have at present not found normalised accident data for UTO (incidents based on person km), and no accidents have been reported. We have found reports in newspapers about minor incidents, without any fatalities reported. Based on data and experiences so far, it seems that the UTO has exceptionally high safety. However, more systematic analysis and normalisation of all international UTO transport incidents are needed.

Even though driverless trains have an impressive safety record, experience shows that they still face some challenges related to reliability and operability. One example of this is seen in Singapore. UTOs were introduced in Singapore's Mass Rapid Transits (MRT) system in 2003. Here, the operations were monitored remotely from an operations control centre. However, in 2018, most of these trains were manned again, for improving reliability. Some of the trains experienced technical issues and failures. In these cases, a driver on board a train will immediately be able to assess the problem, and, if necessary, push another disabled train out of the way. With a driverless system, a driver had to make his way to the unmanned train, which takes time. Nevertheless, the safety record of driverless trains is impressive, maybe due to the rail track as a system. Hence, further automation of railway systems is ongoing.

Lesson Learned

As mentioned, it seems that the UTO has an exceptionally high level of safety. However, systematic analysis and normalisation of all international UTO transport incidents are needed. Thus, there is a need for systematic reporting and analysis of minor incidents/small accidents in order to support risk-based regulation and risk-based design of the technology.

A key issue related to safety is the focus on a restricted design domain and operational envelope. The environment/context of which the UTOs operates is typically underground, with few or no interaction with other traffic. Protection systems are in place at the embarkment area/platform preventing the most common incidents (people falling on tracks). There has been a focus on analysing personnel incidents when entering and leaving the UTOs and building safer infrastructure to minimise dangerous situations.

Autonomy on Road

Cities worldwide are increasingly testing and implementing autonomy as the pace of autonomous vehicle innovation picks up. Norway has long-term experiences of autonomous transport systems such as Automated Guided Vehicles (AGVs) at St. Olav Hospital and autonomous shuttle buses used from January 2018 on public roads.

Projects with autonomous vehicles (AVs): Local governments must approve self-driving pilots. In the US, in California, all companies must deliver annual self-reports on incidents with highly automated vehicles. (This is one of the reasons why Uber and many other companies moved the testing of self-driving taxis to Arizona that has adopted a more liberal attitude.) This framework condition, i.e. legislation in California, has enabled the industry to document the level of safety and identify challenges.

Related to the present development trends, there are two clear trends that are different in nature:

1. a race to develop fully AVs, i.e. self-driving cars, aiming to replace today's private cars.
2. an effort to develop fully AVs to provide mobility-as-a-service (MAAS) or robotaxis.

The aim of the private self-driving car segment is to operate more safely than human drivers are able to in real-world conditions and at high speed. Here, the self-driving cars must be able to handle all types of obstacles and interactions with other road users in all kinds of weather and traffic conditions.

The MAAS segment focusses on small shuttle buses (or robotaxis) with geofencing to establish a safe route. Many of these are unable to go around an obstacle. They stop until the obstacle has moved or been removed. They operate at low speeds between 12 and 30 km/h.

There are many projects with self-driving vehicles on public roads operating around the world. According to Philantropies (2017), at least 53 cities are currently involved in testing AVs. Legal frameworks for the regulation of pilot testing are established in Singapore, the Netherlands, Norway and the UK (KMPG, 2018). Euro NCAP has designed a set of test procedures for testing automated vehicles on SAE level 2. The US Department of Transportation has developed a framework (NHTSA, 2018) for testing automated driving systems focussing on failure behaviour, failure mitigation strategies and fail-safe mechanisms.

AGVs at St. Olav Hospital have been in operation since 2006. Today, 21 AGVs operate at a speed of approximately 2 km/h (max speed is 5 km/h) and communicate with each other, open doors and reserve elevators. The automation is quite simple as they follow a predefined path, and when there are conflicts or problems with collisions/doors/ elevators, a signal is given to the operational centre, always manned by an operator who can intervene or go to the place. Manned operators in the centre are necessary to ensure continuous operations. Even in this strict operational envelope, humans are critical components in the loop. Sensemaking has been in focus, examples are that the AGVs are "speaking" to hindrances/people – saying "please move" or "this elevator is reserved".

Pilots with autonomous shuttle buses: From 2017, testing of AVs was allowed in Norway. In the SmartFeeder (2019) research project, initial data are gathered from five test sites with MAAS pilots. Each pilot tests self-driving shuttle buses carrying up to six passengers, operating at an average speed of 15 km/h, and with an operator to monitor and take over control if necessary (during the test phase). These pilots are "fixed route autonomy", where the autonomous system follows a predefined route and processes a limited amount of sensor data along the route. The motivation varies, i.e. solving a last mile problem (connecting workplaces with public transportation), testing out technology and user acceptance or property and business development. In total, the buses in the pilots have driven almost 22,000 km, with approximately 40,500 passengers in both summer and winter conditions. Initial data have been collected regarding disengagement of the system and involvement of the operator in the pilots in three categories: "obstacle emergency stop" (sensors detect something and automatically stop), "soft stop" (operator overtakes system and decelerates the vehicle) and "Manual switch" (for manually driving the vehicle). The collected data are currently being processed and cleaned for more detailed analysis, and interpretations cannot be drawn yet. However, the reliability and robustness are challenging, and demands a restricted operating envelope in addition to the need for "humans in the loop" when the unanticipated is happening.

Safety Challenges

Tesla with its autopilot has enabled automated driving at high speeds. Several severe accidents with Tesla autopilot have led Tesla to limit their autopilot functionality. These partially automated vehicle systems at SAE level 2 (SAE, 2018) always operate exclusively based on an attentive driver being able to control the vehicle. For fully automated driving (SAE level 4–5), the driver is no longer available as a backup for the technical limits and failures. Replacing human action and responsibility with automation raises questions of technical, ethical and legal risks, as well as product safety.

As far as we know from media and public accident reports there have been four fatal accidents worldwide: three with semi-automated (SAE level 2) autopilot and one with a more fully automated vehicle on public roads (SAE level 3), the Uber accident in Arizona where a Volvo refitted with Uber self-driving technology killed a pedestrian (NTSB, 2018). In all cases, the autopilot was engaged but without driver interaction or intervention with vehicle controls, highlighting the need for sensemaking and "meaningful human control".

There are few safety records (data) on SAE level 4 so far. Data from 2009 to the end of 2015 collected by Google's cars list three police reportable accidents in California while driving at 2,208,199 km (Teoh and Kidd, 2017). This is 1/3 of reportable accidents per km of human-driven passenger vehicles in the same area. In 2017, 19 of 21 reported accidents with Google-Waymo cars (level 4) were rear-ended accidents at signalised intersections. This is caused by ordinary drivers' misinterpretation of automated vehicle behaviour (as an example expecting that drivers are not halting when meeting a yellow light at an intersection.). Google-Waymo has now patented a software program allowing their vehicles to drive through yellow light. A look at accidents and incidents reported to the California Department of Motor

Vehicles (DMV) in 2019 shows that other 65 companies currently testing level 4 technology still have frequent rear-end collisions at signalised junctions. They also have trouble (and reported accidents) entering a motorway from the ramp. AVs have not yet learned the "nudging" that ordinary drivers do to see if traffic on the motorway yield and let you in.

Experience from the autonomous shuttle buses: For the pilots, it was mandatory to report incidents and accidents. No persons were injured, and only minor technical issues and malfunctions were reported. The following issues were revealed:

– Snow, heavy rainfall and fog are challenging for the sensors.
– Vegetation and light poles along the route of the bus is challenging as they interfere and disturb the sensors at times.
– The buses run along the same "track" with narrow wheels, causing significant wear and tear on the road along this track.
– Cyclists passing near the bus makes the bus stop abruptly.

These issues are related to the predefined operational envelope surrounding the vehicle, leading to abrupt stops when violated. As pointed out by Jenssen et al. (2019), AVs lack a sense of self, and software and sensors are still not designed to account for the discrepancy in the same way human drivers are able to.

When applying for testing, a mandatory risk assessment was carried out. The main risks listed were related to passenger injury as a result of an abrupt stop where passengers inside the bus are unprepared and can be harmed by falling. Risk-reducing measures are lowering the speed, installing seat belts, limiting the number of passengers and adding road signs.

AGVs at St Olav: A total of 100–130 minor incidents per year have been reported. Yearly, each AGV experiences around 15 emergency stops (Johnsen et al. 2019), where components must be changed. Reported incidents are minor crashes as a consequence of faulty navigation due to objects placed in the route, summarised in Johnsen et al. (2019). From interviews with the operators of the AGVs, the following main issues are identified:

– The AGVs ability to adapt to the surrounding infrastructure
– Keep the track of the AGVs clear of objects
– Make objects visible to the AGV: the AGVs are not able to detect all obstacles due to the sensor range
– Establish a control room with proper HMI design
– Maintain the interface to cyber physical systems: software updates has led to problems (due to poor testing and multiple vendors.)

Lessons Learned

Vehicle automation can enhance safety but also introduces new risks due to poor technical implementation and the need for rapid response from the human actor. This is especially the case with SAE automation levels 2 and 3.

The accident data collected so far with automation (AGVs and level 1–4 vehicles) indicate safety hazards of human factors and technical issues, i.e. obstacle detection (sensors), programming (rule-based and not artificial intelligence, AI), prolonged attention (humans in the loop), HMI (Autopilot-engagement rules) and misuse. The list may become longer as more safety data are gathered and more in-depth information on accident causality of automated vehicles is established, e.g. overreliance and expectation mismatch.

Based on the experiences, there is a need to establish regulations that ensure systematic incident reporting, develop systems based on learning from incidents and invest in infrastructure to support automation, i.e. help the automation by focussing on an operational envelope that uses more data from infrastructure. The transport systems are automated but not autonomous. Autonomous systems are immature at present and must be further developed.

A SUMMARY OF MTO SAFETY ISSUES

Based on the performed reviews, the suggested key measures are listed below.

Humans: As seen from all experiences, the uncertain and complex environment for autonomous systems must ensure the need for human intervention. Autonomous transportation systems will to a varying degree need human control if failures occur or under certain operational conditions. With today's UTOs and AGVs, an operator is still needed when there is a disruption and sensors fail to detect and recognise an obstacle or determine the next actions. However, in testing and developing autonomous transportation systems with drones, AVs and vessels, we see examples of projects where the human operator is not considered from the beginning. The industries' motivation seems to be to try to automate as much as possible and assume that humans will and can monitor it. Hence, HAI and how to keep the humans in the loop is often considered a challenge to be solved late in the project after knowing the limitations of the technology and by considering the humans as the adapting back-up. Most of the projects lack early incorporation of human factors in analysis, design, testing and certification process. Thus, there are costly challenges that should have been addressed earlier by starting with technology, human limitations and possibilities, and organisational and infrastructure needs. A key issue is to define the design conditions the system should operate under by defining the operational envelope and critical scenarios (such as sensor failures). Then specify how critical scenarios can be mitigated by infrastructure support i.e. surrounding systems such as other autonomous systems nearby (cars) or control infrastructure. If human intervention is needed to handle the scenarios, sense-making must be supported within the existing limitation of human abilities.

As aviation is the industry with the most experience with safe automated systems, the list from Endsley (2019) with design principles for improving people's ability to successfully oversee and interact with automated systems should be a very useful element, allowing for manual overrides and sufficient training to users on automation to ensure adequate understanding and appropriate levels of trust.

Technology: To date, developing autonomous or remotely controlled transportation systems (especially for AVs and MASS) appears to primarily be about a technology push rather than considering and providing sociotechnical solutions including redesign of work, capturing knowledge and addressing human factors as we and others have seen (Lutzhoft et al., 2019).

Technology in autonomous systems and their interpretation (such as through AI) are not reliable at present – thus, there is a need to address poor reliability trough improving man/technology/organisation aspects. The reliability of drones is lower than for manned planes, and we have seen how sensors and technical equipment are causing safety issues in several projects. The systems must improve for an industrial setting and for safety-critical operations, i.e. become highly reliable and resilient to bad data and have automatic self-checking behaviour and avoiding single-point failures by checking across multiple inputs. Thus, there is a need to get support from other AVs with sensors, need for developing infrastructure (such as roads and seaways with sensors), in addition to establishment of control centres for road traffic and maritime traffic that must be responsible for supporting sensemaking among the actors (i.e. automated and not automated systems). Technical barriers must be in place to a larger extent on autonomous systems to avoid and reduce the outcome of failures and component interaction accidents, which are more common as the complexity increases.

Automation transparency is important for both sharing the situation awareness and communicating the intentions towards others and for the operator in an RCC to understand the behaviour of the automation. In complex systems, a wide range of alarm issues related to diagnostics, management and assessments of multiple input data will be challenging. Hence, alarms must be unambiguous and displayed with a clear message. This requires good human factor engineering practice, such as an alarm philosophy and relevant standards.

Organisation: Experience from the projects and pilots demonstrate a need to see the technological solution in a larger sociotechnical context. Autonomous transportation systems are a system of systems. We have seen that legislation is are needed to gather data and establish the operational context. There is a need for substantial investments in infrastructure: organisational interfaces are lacking and organisational/structural issues from the operator/company/area/society are often considered the last thing to get in place. Looking at the operational context, we have seen a need to limit the operational design domain and use operational envelopes, or safety envelopes to define situations, responsibilities and system characteristics during all conditions (especially in safety-critical conditions with sensor/data failures). Regulations and guidelines have slowly been established to support autonomous transportation systems. However, few of them require systematic reporting of accidents and incidents. Experience from accidents with AVs has given valuable insight, and hence all domains should prioritise and require reporting and systematic data collection of failures, hazards and unforeseen events. Not requiring reporting and sharing of safety-critical systems is a risk in itself.

SENSEMAKING TO SUPPORT MEANINGFUL HUMAN CONTROL

Focus on the design of operational envelopes to reduce complexity and analysing the needs for cues and information to support sensemaking and meaningful human control, when needed, is a key issue. Defining operational envelopes answers the question of which functions and roles automation/autonomy should have, versus humans, when designing a complex system. This is also an important question for certification of the autonomous transportation system.

Sensemaking and the principle of meaningful human control should be used to verify that the proper functions are allocated to the human or the automation. According to Santoni de Sio and van der Hoven (2019), two design requirements should be satisfied for an autonomous system to remain under meaningful human control:

1. A "tracing" condition, according to which the system should be designed in such a way as to grant the possibility to always trace back the outcome of its operations to at least one human along the chain of design and operation.
2. A "tracking" condition, according to which the system should be able to respond to both the relevant moral reasons of the humans designing and deploying the system and the relevant facts in the environment in which the system operates.

From a safety perspective, this can be placed in the bowtie model, where the design principle of tracking are barriers preventing a technical fault, threat or unexpected situation to lead to a dangerous situation, as a human alway has established the possibility to intervene and take over control. On the other side of the bowtie, once a hazard has emerged, the outcome can be reduced by designing after a tracing condition making it possible to trace back the operation to a human who is in the position to understand the capabilities of the system and the possible effects in the world of its use and, hence, knows how to limit the consequences of an undesired event.

CONCLUSION

We have given a summary of ongoing projects and safety issues. The main issues across the domains are technical reliability and maturity, the need for automation transparency (including awareness for the decision made by automation), the need for defining what conditions the system can operate under and assigning responsibilities to human operators and the automation. Experiences from known accidents involving a high level of automation, as in the cases of Boeing 737 MAX, Uber and Tesla, have shown overreliance on automation and poor understanding of capabilities and limitations. We need to collect and systemise data on accidents and incidents of autonomous transportation systems and design with human factor practice to support sensemaking and meaningful human control.

Design principles from meaningful human control should be used to verify if the interaction between automation and the human is safe. This can be used as an input to operational envelopes and to assist in the design of a good HAI supporting sensemaking.

ACKNOWLEDGEMENT

This chapter has been funded by the Norwegian Research Council – project 267860 SAREPTA.

REFERENCES

AAWA (2020). https://www.rolls-royce.com/media/press-releases/2016/pr-12-04-2016-aawa-project-introduces-projects-first-commercial-operators.aspx

AMOS (2020). https://www.ntnu.edu/amos/research

Autosea (2020). https://www.ntnu.edu/autosea

Autoship (2020). https://www.kongsberg.com/maritime/about-us/news-and-media/news-archive/2020/autoship-programme/

Bureau Veritas (2019). NI 641 R01 *Guidelines for Smart Shipping.*

Chinen, M. (2019). Law and Autonomous Machines. *Elgar Law, Technology and Society* (p. 109). Edward Elgar Publishing.

Cruz, B. S., & de Oliveira Dias, M. (2020). Crashed Boeing 737-MAX: Fatalities or malpractice? *GSJ* 8 (1), 2615–2624.

Cummings, M. L. (2019). *Lethal Autonomous Weapons: Meaningful human control or meaningful human certification?* IEEE Technology and Society.

Endsley, M.R. (2019). *Human Factors & Aviation Safety* Testimony to the United States House of Representatives. Hearing on Boeing 737-Max8 Crashes, December 11, 2019.

Fjørtoft, K. E., & Rødseth, Ø. J. (2020). *Using the operational envelope to make autonomous ships safer* Proceedings of the 30th European Safety and Reliability Conference and the 15th Probabilistic Safety Assessment and Management Conference Edited by Piero Baraldi, Francesco Di Maio and Enrico Zio.

Hoem, Å. S. (2019). The present and future of risk assessment of MASS: a literature review. *29th European Safety and Reliability Conference.* European Safety and Reliability Association.

Hoem, Å.S., Fjørtoft, K., & Rødseth, Ø. (2019): *TransNAV 2019: Addressing the Accidental Risks of Maritime Transportation: Could Autonomous Shipping Technology Improve the Statistics?*

Hollnagel, E., Nemeth, C. P., & Dekker, S. (Eds.). (2008). *Resilience engineering Perspectives: Remaining Sensitive to the Possibility of Failure* (Vol. 1). Ashgate Publishing, Ltd.

Horowitz, M., & Scharre, P. (2015). *An Introduction to Autonomy in Weapon Systems.* Center for a New American Security (CNAS) Working Paper (CNAS: Washington, DC), p. 8

IMAT (2020). https://www.sintef.no/projectweb/imat/

INAS (2020). http://www.autonomous-ship.org/index.html#H2

Johnsen, S. O., Hoem, Å., Jenssen, G., & Moen, T. (2019). Experiences of main risks and mitigation in autonomous transport systems. *Journal of Physics: Conference Series* 1357 (1) 012012.

Kilskar, S. S., Danielsen, B. E., & Johnsen, S. O. (2020). Sensemaking in critical situations and in relation to resilience—a review. *ASCE-ASME Journal of Risk and Uncertainty in Engineering Systems, Part B: Mechanical Engineering,* 6(1).

KMPG (2018). Autonomous vehicles readiness index. *Klynveld Peat Marwick Goerdeler (KPMG) International.*

Lutzhoft, M., Hynnekleiv, A., Earthy, J. V., & Petersen, E. S. (2019). Human-centred maritime autonomy-An ethnography of the future. *Journal of Physics: Conference Series* 1357 (1), 012032.

MUNIN (2020). http://www.unmanned-ship.org/munin/

NFAS (2020). http://nfas.autonomous-ship.org/index.html

NHTSA (2018). *A Framework for Automated Driving System Testable Cases and Scenarios.* DOT HS 812 623. https://www.nhtsa.gov/sites/nhtsa.dot.gov/files/documents/13882-automateddrivingsystems_092618_v1a_tag.pdf

NTSB (2017). National Transportation Safety Board 2017. Collision between a Car Operating With Automated Vehicle Control Systems and a Tractor-Semitrailer Truck Near Williston, Florida, May 7, 2016. Highway Accident Report NTSB/HAR-17/02. Washington, DC.

NTSB (2018). *National Transportation Safety Board 2018*. Preliminary Report: Highway HWY18MH010.

Porathe, T. (2019). Interaction between Manned and Autonomous Ships: Automation Transparency. *Proceedings of the 1st International Conference on Maritime Autonomous Surface Ships*.

Porathe, T., Hoem, Å., Rødseth, Ø. J., Fjørtoft, K., & Johnsen, S.O. (2018). At least as Safe as Manned Shipping? Autonomous Shipping, Safety and "Human Error". *Proceedings of ESREL 2018*, June 17–21, 2018, Trondheim, Norway.

Ramos, M. A., Utne, I. B., Vinnem, J. E., & Mosleh, A. (2018). Accounting for Human Failure in Autonomous Ship Operations. Safety and Reliability–Safe Societies in a Changing World. *Proceedings of ESREL 2018*, June 17–21, 2018, Trondheim, Norway.

Relling, T., Lützhöft, M., Ostnes, R., & Hildre, H. P. (2018). A Human Perspective on Maritime Autonomy. *International Conference on Augmented Cognition* (pp. 350–362). Springer, Cham.

Rødseth, Ø. J. (2018). Defining Ship Autonomy by Characteristic Factors, *Proceedings of ICMASS 2019*, Busan, Korea, ISSN 2387–4287.

SAE International (2018). Standard, SAE J3016_201806. *Taxonomy and Definitions for Terms Related to Driving Automation Systems for On-Road Motor Vehicles*. Revised

Santoni de Sio, F., & Van den Hoven, J. (2018). *Meaningful human control over autonomous systems: a philosophical account. Frontiers in Robotics and AI* 5, 15.

SmartFeeder (2019). https://www.sintef.no/prosjekter/smart-feeder/ (in Norwegian)

Teoh, E. R., & Kidd, D. G. (2017). Rage against the machine? Google's self-driving cars versus human drivers. *Journal of Safety Research* 63, 57–60.

Thieme, C. A., Guo, C., Utne, I. B., & Haugen, S. (2019, October). Preliminary Hazard Analysis of a Small Harbour Passenger Ferry–Results, Challenges and Further Work. *Journal of Physics: Conference Series* 1357 (1), 012024).

UITP (2013). *Observatory of Automated Metros World Atlas Report*. International Association of Public Transport (UITP), Brussels

Wang, Y., Zhang, M., Ma, J., & Zhou, X. (2016). Survey on driverless train operation for urban rail transit systems. *Urban Rail Transit* 2, 106–113. https://doi.org/10.1007/s40864-016-0047-8

Yara Birkeland (2020). https://www.kongsberg.com/maritime/support/themes/autonomous-ship-project-key-facts-about-yara-birkeland/

Zeabuz (2020). https://zeabuz.com/

13 Application of Sensemaking

*Data/Frame Model, to UAS
AIB Reports Can Increase
UAS GCS Resilience
to Human Factor and
Ergonomics (HF/E) Shortfalls*

Q. R. Waraich
The George Washington University

CONTENTS

INTRODUCTION

There are more than 10,000 Unmanned Aerial Systems (UAS) in the air around the world at any given time (Waraich, Mazzuchi, Shahram, & Rico, 2013). Over the past two decades, UAS have grown exponentially (Matolak & Sun, 2015). Lack of onboard pilots meant that UAS can be sent deep into hostile environments without having to fear for pilot safety (Agarwal, Murphy, & Adams, 2014). It is an appealing aspect for security, law enforcement, and military for conducting intelligence, surveillance, reconnaissance, search, and rescue (Gawron, 1998). Its pilot safety feature initially helped steer the UAS developmental focus mainly towards military use. To meet the warfighters demands, military UAS were hastily developed and deployed, thus leading to an increased number of UAS mishaps (Baur, 2007; Nisser & Westin, 2006).

Nevertheless, the same pilot safety feature led to bypassing majority of the standardized testing that was conducted for manned fighter aircrafts prior to their deployment. Nowadays, the use of UAS has expanded in all sectors. The UAS are assisting to perform all sorts of dangerous and dirty civilian tasks while gathering high technical quality data. Nonetheless, similar weaknesses in testing of industrial UAS systems persist.

Several UAS mishap studies have shown human factors involvement in up to 69% of all such UAS mishaps, and up to 25% are due to ergonomic shortfalls that are found in human–machine interface (HMI) design and configuration of ground control stations (GCS) (Peter & Karl, 2016; Hobbs & Shively, 2014; Williams, 2004; Manning, Rash, LeDuc, Noback, & McKeon, 2004; Thompson & Tvaryanas, 2008; Rogers, Palmer, Chitwood, & Hover, 2004). The design and development of UAS GCS lack HMI-specific human factors and ergonomic (HF/E) standards, leading to varying GCS designs and/or configurations that do not suit the operator (Waraich, Mazzuchi, Shahram, & Rico, 2013). Lack of UAS-specific HF/E standards may have led to following shortfalls in the safety-critical GCS designs: such as, visual/audio information presented in text, complicated sequence of menu selection, unguarded placement of safety-critical controls in areas where they could inadvertently be activated, controls whose functions can be altered by a change in selected mode, out of reach control placement, pop-up windows blocking critical parts of display, and proliferation of screen displays (Hobbs & Lyall, 2016).

As per the studies, seemingly high number of HF/E GCS-related UAS mishaps call for an imminent need to develop an HF/E-specific UAS GCS standard. Generally, a standard takes years to develop. Nevertheless, there are several national and international human factors standards that may have sections related to HF/E. These could possibly be applied to UAS GCS to increase their HF/E resilience. These standards include American National Standards Institute (ANSI)/HFES-100 for human factors Engineering of computer workstation, NASA-STD-3000 for Man-System Integration, ISO 6385 for Ergonomic Design of Control Systems (that sets out the broad principles of ergonomics), ISO 26800 Ergonomics, ISO 11064 for Ergonomic Design of Control Centres, ISA SP101 for Human–Machine Interface, and ISO 9241 covering ergonomics of human–computer interaction.

Similarly, the UAS mishap investigation lacks proper unmanned specific tools (and sometimes competencies) to uncover HF/E issues in the UAS GCS input/output (IO) interface to provide a long-lasting solution by eliminating HF/E issues and adding resilience to its design. Therefore, the manned aircraft investigation taxonomies/models are being used to prepare the Accident Investigation Board (AIB) reports for UAS mishaps.

Over the years, a large majority of UAS GCS are looking more and more like the computer workstation (CW) setup; therefore, the application of ANSI/HFES-100 may help reduce the HF/E-related mishaps (Waraich, Mazzuchi, Shahram, & Rico, 2013). This study was conducted more than 7 years ago, but there continues to be some complacency related to HF/E issues that continue to persist. If GCS are resembling CW, then HF/E standards for CW (i.e., ANSI/HFES-100) shall be applicable to the UAS GCS.

This study takes a two-part approach: first, to reconfirm whether ANSI/HFES-100 still applies and second to explore the possibility of applying sensemaking to uncover the hidden HF/E listed in the AIB reports, along with the application of ANSI/-HFES-100 to mitigate such HF/E.

The term sensemaking refers to simply making sense of the situation at hand, both in terms of system expectations and reality. Sensemaking allows viewing human error as a symptom of the problem rather than being the cause (Kilskar, Danielsen, & Johnsen, 2020). It may help improve the focus on HF/E in control of UAS (through GCS) and add the much-needed resilience to the UAS and GCS design.

BACKGROUND

UAS History

Majority of human history is marred with conflicts. Gaining superiority over one's enemy led to research for such superior weapon. The oldest know unmanned aerial penetration had been attempted more than a century prior to the infamous Wright Brother's aircraft flight at Kitty Hawk, with varying levels of ground control.

- In 1806, Lord Thomas Cochrane flew kites from the decks of Royal Navy frigate HMS Pallas (1757–1783) to drop propaganda leaflets, no ground controls were used.
- In 1849, General Uchatius bombarded the city of Venice, Italy, with balloon bombs, no ground controls were used.
- In 1898, the unmanned systems control mechanism improved substantially when Nikola Tesla wirelessly controlled a small boat (Miessner, 1916; Tesla, 1898)

- In 1912, A. J. Roberts' wireless control of an airship is the oldest known use of wireless technology to control a flying object from ground (Miessner, 1916).
- In 1918, Charles Kettering developed a gyroscope-controlled autonomous flying object (missile) with no ground control mechanism (Brittain, 2009).
- In the early 1920s, the inventor Elmer Sperry used wireless controls to feed path correction to an aircraft for aerial delivery of messages.
- In 1939, hobbyist Reginald Deny built radio control decoy planes for the military for target practice.
- In 1962, a project was started to produce autonomous spy planes, such as the D-21 Tagboard and the Ryan Model 147 Lightning Bug, no over the horizon ground controls were implemented in the design.
- In 1982, the success of Israel's unmanned decoys over Lebanon's Bekaa Valley brought the U.S. military's focus back to the subject of unmanned drones, at the time relays were used to provide over the horizon control of a drone and formal GCS were being used.
- In mid-1990s, the development of GCS took a leap into the future when a formal GCS was developed to control a Predator UAS via satellite (Haines, 2007). In 1999, UAS garnered attention during military operations in Kosovo, when they were mainly used for reconnaissance.

A simplified timeline of UAS development is shown in Figure 13.1.

UAS Development

At the turn of the 21st century, operating UAS controls via satellite from a remote GCS was still a relatively new concept. The GCS design and testing were a lackluster process that was missing adherence to any applicable standards. The developers had not yet decided on principles or paradigms for layout of the safety-critical GCS. The GCS could look exactly like a cockpit or it could simply look like a control room environment. Therefore, a tug of war ensued between these two competing paradigms to operate UAS from the GCS while allowing developers leeway to experiment with varying levels/types of UAS control mechanism. The developers designed new systems to find the right balance between "ground control" and "autonomous controls" (Zhang, Feltner, Shirley, Kaber, & Neubert, 2020; Mouloua, Gilson, Kring, & Hancock, 2001). Figure 13.2 shows both paradigms:

Development of new UAS systems turned into a competition for developers to showcase their latest and greatest UAS GCS technology. This coupled with the lack of applicable HF/E standards for GCS compounded the HF/E shortfalls in UAS GCS, leading to an increased number of UAS mishaps (Waraich, Mazzuchi, Shahram, & Rico, 2013; Nisser & Westin, 2006). Nowadays, UAS have found their purpose in both military and civilian sector. Farmers are utilizing the UAS for crop assessment, forecasting, disease/weed detection, and other applications including fishery, land surveys, oceanography, and firefighting (Berni, Zarco-Tejada, Suarez, & Fereres, 2009; Pastor, Lopez, & Royo, 2007). The hobbyists, photographers, recreationists, and surveyors all have a UAS specifically designed to cater their needs (Hamilton & Stephenson, 2016).

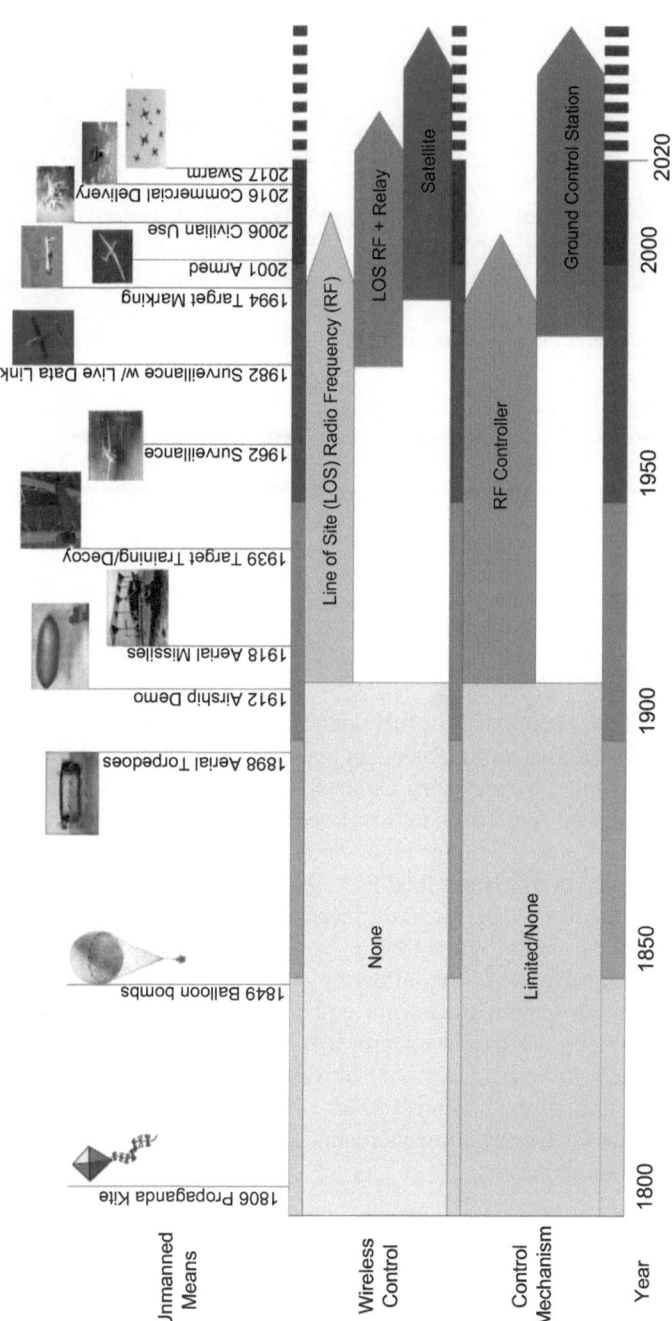

FIGURE 13.1 Timeline of UAS development.

FIGURE 13.2 Cockpit paradigm (left), computer workstation paradigm (right) for GCS.

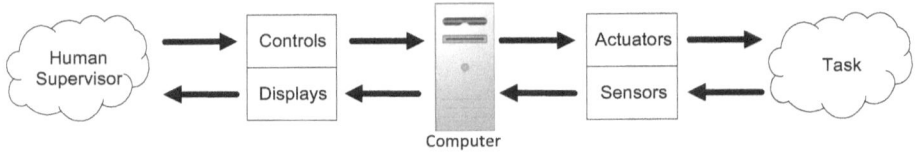

FIGURE 13.3 Human supervisory control. (Adapted from Sheridan, 1992.)

UAS CONTROL MECHANISMS

The GCS provides the means to gain full operational control of the UAS. The UAS GCS can be seen as a remote extension of a cockpit (Guidance, 2008). Sheridan (1992) provided a Human Supervisory Control (HSC) model as an overview of the human–GCS interaction. According to this model, the human interacts with the controls to perform a task; the computer then sends the control data to the machine to perform task and receives machines feedback; that feedback is then sent/displayed to the human as a status of task performed (Sheridan, 1992). All UAS GCS use some level of HSC model shown in Figure 13.3.

Technological improvements are allowing the UAS to be more autonomous. Nevertheless, being autonomous does not equate to completely removing humans from UAS control loop as operators are still needed to perform safety-critical tasks. There are three different UAS control mechanisms being used: autonomous, semi-autonomous, and ground control (Nas, 2008; Mouloua, Gilson, Kring, & Hancock, 2001). The UAS control mechanisms are described below from highest to lowest level of automation (LoA):

- **Autonomous:** UAS are capable of flying a complete mission from takeoff to landing. Operators are only needed to perform supervisory tasks in the GCS. They have the option to adjust the mission plan, but they have no direct control of the UAS.
- **Semi-autonomous:** UAS are capable of executing autonomous tasks. The operators are only needed to perform supervisory tasks in the GCS.

However, they have the moderate level of direct flight control through a joystick.

- **Ground control:** UAS cannot perform mission plan autonomously. The operators are bound to directly control the UAS typically with a joystick from takeoff to landing.

UAS CATEGORIES

The UAS come in different shapes and sizes. The United States Department of Defense (U.S. DoD) has categorized them into five specific groups. The categories are based on altitude, weight, endurance, and speed. Low endurance UAS belong to Group 1 while the high endurance UAS belong to Group 5 (Weatherington, 2010). Table 13.1 describes the UAS groups:

UAS MISHAP STUDIES

Numerous studies indicate that human factors are found to be involved in 50%–69% of all UAS mishaps, and overall 16%–25% of these mishaps are directly associated with HF/E in UAS GCS designs. The variation between the results of these studies is due to the varying taxonomies/models used when performing analyses. Below is a snapshot of these UAS mishap studies:

- Manning applied the human factors analysis and classification system (HFACS) model to assess the UAS mishaps. He found that 61% of mishaps studied were due to human factors; of these 19% were linked to HF/E in UAS GCS. Study highlighted inadequate attention to ergonomics of control layouts and input devices as being the probable cause for high HF mishap rate (e.g., wrist bent angle when using GCS IO, position/location of IO controls) (Manning, Rash, LeDuc, Noback, & McKeon, 2004)
- Rogers applied human-system issues taxonomy to study human factors in UAS mishaps. He found that 69% of mishaps studied were due to human factors; of these 16% were linked to HF/E in UAS GCS. He found that, HF/E-related mishaps can be minimized if the UAS developers concentrate on human-system integration of GCS IO devices when designing and testing

TABLE 13.1
UAS Groups (Chanda, et al., 2010)

UAS	Weight (lb)	Altitude	Air Speed (knots)
Group 1	0–20	<1,200 AGL	<100
Group 2	21–55	<3,500 AGL	<250
Group 3	<1,320	<18,000 MSL	
Group 4	>1,320		Any airspeed
Group 5		>18,000 MSL	

(e.g., display, joystick, keyboard, mouse) (Rogers, Palmer, Chitwood, & Hover, 2004; Williams, 2004).

- Thompson applied (Non-DoD) HFACS model to UAS mishap reports. He found that 60% of all mishaps studied were due to human factors; of these 25% were ascribed to the HF/E in UAS GCS. His recommendations include optimization of UAS GCS interface by refining the layout of IO devices to provide an ergonomic fit for operators (e.g., display, keyboard, mouse, joystick) (Thompson & Tvaryanas, 2005).

- Thompson applied the DoD HFACS model to study UAS mishaps. He found that 50% of mishaps were related to human factors; of those 19% were ascribed to HF/E in UAS GCS. His findings included improper clearance and alignment of IO controls (e.g., display, keyboard, joystick, mouse) to be the leading cause in HF/E-related UAS mishaps (Thompson & Tvaryanas, 2008).

UAS MISHAP INVESTIGATION MODELS FOR HF/E EVALUATION

If the human developers/operators of technology-intensive systems see deficient or unsafe process or a system, they would not allow it to continue (Dekker, 2001). However, they do not foresee the accidents, because they cannot predict the possibility of accident actually happening (Wagenaar & Groeneweg, 1987). As mentioned by Dekker (2004) human error can be seen as a consequence of deeper issues (i.e., poor design, poor training, mental overload, fatigue) with the system (Dekker, 2004). In safety-critical GCS, accidents typically have more than one reason, as they are a collection of knottily interrelated chain of events. Where these knots need to be untied delicately to see the actual flow of events leading to an accident, an accurate and comprehensive post-accident documentation is the most important step that can result in improving the resilience and safety of the system and possibly halt similar recurrences (Gordon, Jeffries, & Flin, 2002), preferably with the involvement of human factors experts.

Therefore, taxonomies/models are continually being developed and improved by researchers to methodically collect and document post-mishap data for a human factors analysis. Since, varying perspectives on human factors are applied during the development of such models, it is nearly impossible to select an appropriate model for capturing HF/E-related mishap data, because not all models can properly evaluate HF/E issues (Andersen et al., 2002; Wiegmann & Shappell, 2001). On the other hand, some models may be good at capturing IO-related HF/E data during a mishap investigation, but may not be able to properly evaluate collected data to find the underlying HF/E issues, leading to a missed opportunity to find and eliminate HF/E issues retrospectively. A report by the European Organization for the Safety of Air Navigation provided an overview of models used in aviation mishap investigations (Andersen et al., 2002):

- **Task-based taxonomies:** Captures data from the operator's perspective. IO HF/E issues can only be found if an operator realized and highlighted such shortfalls.

- **System-oriented taxonomies:** Captures data from the system's perspective. IO HF/E issues may or may not be captured.
- **Communication system models:** Captures data from communications perspective (i.e., messages sent/received and interpretation). Does not evaluate IO HF/E issues.
- **Information processing models:** Captures data by measuring operator's memory, judgment, and decision-making with respect to actions performed. Could possibly capture IO HF/E issues.
- **Symbolic processing models:** Captures data from human thought processes and perspectives. May not capture IO HF/E issues.
- **Situation awareness models:** Studies the information available to the operator prior to decision-making. Could possibly capture IO-related HF/E issues.
- **Control system models:** Evaluates theoretical performance in a closed-loop system, situation, or scenario. May not capture IO-related HF/E issues.
- **Error of commission models:** Evaluates operators competences with respect to the unintended/unnecessary acts performed before mishap. Could possibly capture IO HF/E issues.
- **Human-system issues taxonomies:** Evaluates human-system integration relative to physical, interpretational, and decision-making ability. Could possibly capture IO HF/E issues.
- **Air traffic management models:** Evaluates the cause of incident and studies its relationship with other possible causes. Could possibly capture IO HF/E issues.
- **HFACS models:** Evaluates four causation levels: organizational influence, unsafe supervision, preconditions related to unsafe acts, and unsafe acts. Could possibly capture IO HF/E issues.
- **DoD HFACS models:** This model is an enhancement of HFACS model. Could possibly capture IO HF/E issues.

APPLIED STANDARDS AND MODELS

ANSI/HFES-100 Standard

Individualized CWs have been around for over 30 years. The Human Factors and Ergonomics Society (HFES) is an internationally recognized nonprofit organization, founded in 1957. HFES has an ANSI-approved standard for HF/E in CWs. Their ANSI/HFES-100 standard has been widely accepted by industry since 1988 (ANSI/HFES-100, 2007).

ANSI/HFES-100 lists system/component-level quantitative parameters to assist engineers in designing a system around human limitations and to provide ergonomic CW environment for users. The standard has been applied on CW-based emergency dispatch centers, factories, control rooms, and power plants (ANSI/HFES-100, 2007). The standard has two HF/E categories: operator comfort (OC) and IO. The OC is outside the scope of this study (definition included as reference for future studies). Both IO and OC are described below:

- **Operator Comfort (OC):** OC provides specifications (such as viewing, posture, arm position, work surfaces, and foot comfort) to design an ergonomic CW layout and reduce operator discomfort (Sauter, Schleifer, & Knutson, 1991). OC also provides specifications for lighting, acoustics, temperature, ventilation, and emissions to improve operator well-being and productivity (Corlett & McAtamney, 1988).
- **Input Output (IO):** IO provides HF/E specifications for operator-friendly HMI. IO devices addressed by the standard include keyboard, mouse, trackball, puck device, light pen, stylus, tablet, overlay, touch panel, joystick, and display (ANSI/HFES-100, 2007). Part of this study is designed to validate the applicability of ANSI/HFES-100 standard's IO category on UAS GCS IO interfaces.

SENSEMAKING: DATA/FRAME MODEL

Humans have been applying "sensemaking" since we first started asking questions such as who? when? how? where? and why? The literal definition of the word "sensemaking" is as simple as it sounds; "making sense" of the situation at hand. Sensemaking can also be viewed as a pursuit for accuracy (Gioia, 2006). The concept of sensemaking originated as an organizational literature in 1960s (Maitlis & Christianson, 2014). Sensemaking guides evaluators to assess the situation, with a purpose of extracting meaningful data points (clues). That can explain the situation retrospectively. The main objective of sensemaking is to envisage the relationship between actions performed and the action's interpretation, without being influenced by the availability of choices (Weick, Sutcliffe, & Obstfeld, 2005). Therefore, sensemaking is considered as a process that is prompted by violated expectation (Kilskar, Danielsen, & Johnsen, 2020), thus resulting in leaving cues in the environment. These cues are then collected through an iterative investigation process to find further clues, leading to a clear/better understanding of the situation (Maitlis & Christianson, 2014).

In the context of UAS GCS application, a data/frame model has been researched by United States Army Research Institute for the Behavioral & Social Sciences. This model can be tweaked and utilized to extract valuable information for the purpose of this study. The model was designed to evaluate/analyze situations and be able to put together a puzzle that an untrained eye cannot see (Sieck, Klein, Peluso, Smith, & Harris-Thompson, 2007; Klein, Phillips, Rall, & Peluso, 2007).

The operator's sensemaking must support decisions and continuous adjustments that are made to the UAS, which is flying in a removed, intricate, and rapidly changing environment (Kaste, 2012). With the help of "Data/Frame theory of Sensemaking," we can envision how UAS GCS operators use knowledge to handle complex situations during flight. According to this model, coders (evaluators) review data to find cues and assign a frame to it for organization. As more cues are acquired, the frame is questioned. The coder may restore or replace it with more precise frame. The organized data allow coders to visualize information that was missed or overlooked before (Klein, Phillips, Rall, & Peluso, 2007). The framing process used in this study for sensemaking is derived from the "Data/Frame Model." Figure 13.4 shows a variation of the model intended to be used in this study.

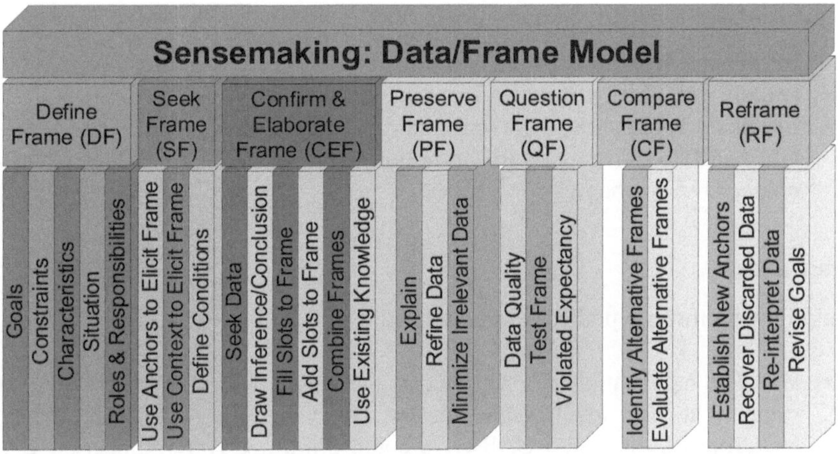

FIGURE 13.4 Sensemaking data/frame model.

RESEARCH METHOD

A mixed-method research design was selected for this study. Many journals and researchers accentuate the application of mixed-method when working with a combination of quantitative and qualitative data (Creswell, 2009). The mixed-methods approach is flexible and allows data analysis/integration at any point during the research (Hanson, Creswell, Clark, Petska, & Creswell, 2005).

BOUNDARIES OF THE STUDY

The UAS, GCS, and human factors engineering fields are very broad. Therefore, this UAS research is limited by the following factors:

- Groups 2 through 5 GCS (see Table 13.1)
- Only fixed GCS (excludes handheld/deployable)
- The study of 36 GCS and 41 operator surveys
- ANSI/HFES-100 IO category
- Sensemaking data/frame model

RESEARCH QUESTION

- Are the IO devices used by the UAS GCS IO interface the same as the ones used in the CW IO interface?
- Is the usage of IO device between CWs and UAS GCS similar?
- Can the application of ANSI/HFES-100 coupled with sensemaking be used to evaluate AIB report to help identify shortfalls in IO design and reduce HF/E issues in UAS GCS?

GOALS AND OBJECTIVES

- Establish similarity between CW and UAS GCS IO devices.
- Establish similarity of IO device usage between CWs and UAS GCS.
- Establish applicability of sensemaking data/frame model and ANSI/-HFES-100 to AIB reports to capture the HF/E in UAS GCS IO interface to reduce UAS mishaps.

HYPOTHESIS

The following three hypotheses were formed:

- **H_1:** The application of sensemaking coupled with "ANSI/HFES-100" standard to help identify and resolve HF/E in GCS IO interface design for Groups 2 through 5 UAS GCS could minimize HF/E impact on UAS operation.
- **H_2:** The "ANSI/HFES-100 2007" standard's IO category applies to Groups 2 through 5 UAS GCS IO interfaces.
- **H_3:** The IO devices used by Groups 2 through 5 of UAS GCS are similar to those listed in the CW "ANSI/HFES-100" standard's IO category.

If both hypotheses H2 and H3 apply, then H1 may also apply. The relationship between research questions and hypotheses is illustrated in Figure 13.5.

RESEARCH DESIGN

The CW IO devices are relatively similar to the ones used in UAS GCS. Waraich (2013) studied 20 GCS and found that 98% of IO devices and their usage in the GCS were similar to that of CWs (Waraich, Mazzuchi, Shahram, & Rico, 2013). The study was limited to U.S. DoD UAS. Since then, great strides have been made in civilian side of UAS industry. Therefore, it is imperative to update 2013 study to reassess its findings and to include UAS from the civilian sectors as well.

To update this study, a research in public domain resulted in 75-candidate UAS from Groups 2 through 5 (2018 RPAS Yearbook, 2018). Further research eliminated 30 of the UAS due to their noncompliance with fixed GCS requirement. This resulted with 45 UAS GCS on a shortlist for this study; program managers, engineers, and operators were contacted to evaluate their willingness to support this study.

This study will be conducted in three phases described below:

Phase I (Quantitative): Identifies IO devices used in study of UAS GCS. The data collected will be utilized in phase II.

Phase II (Quantitative): List of GCS IO devices from phase I, will help generate phase II questioner to help evaluate the similarities of IO devices usage between the GCS and CW. Under both "normal operation" and "emergency operations."

Phase III (Quantitative/Qualitative): Sensemaking data/frame model will be applied both quantitatively and qualitatively to the AIB report's HF/E section. It may help extract information overlooked by the taxonomy used for AIB report

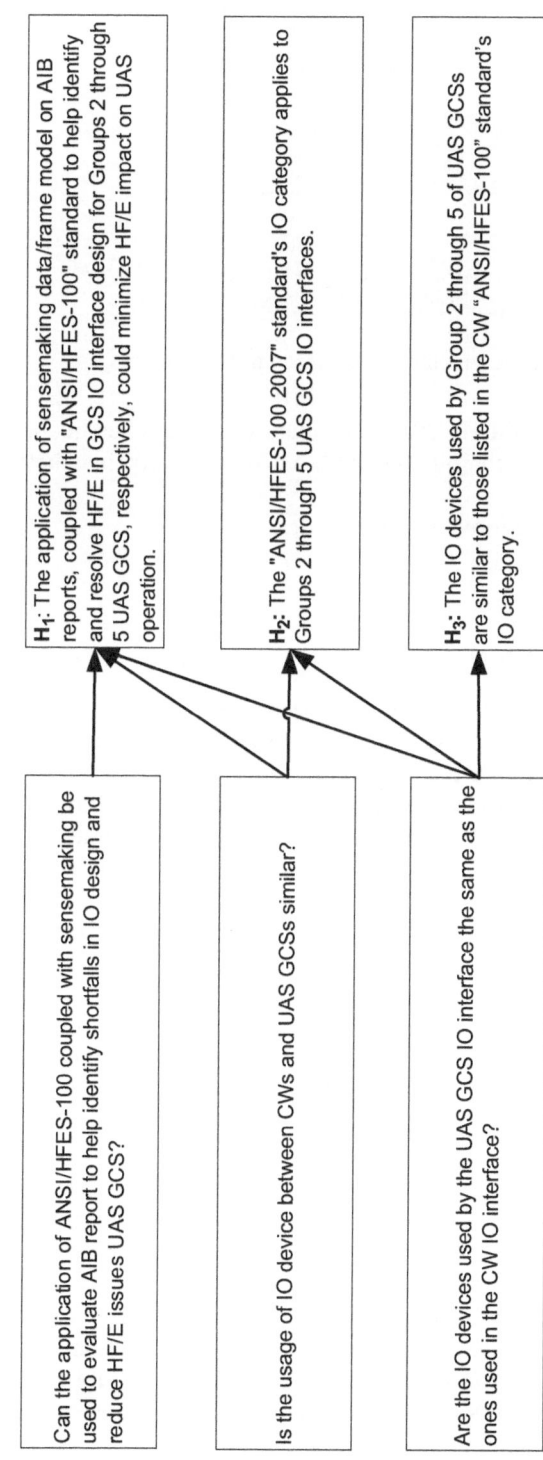

FIGURE 13.5 Relationship between research question and hypotheses.

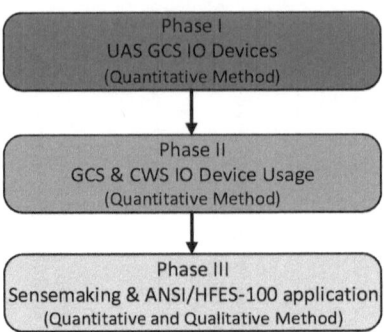

FIGURE 13.6   Three-step research process (mixed methodology).

development. Lastly, ANSI/HFES-100 standard will be applied qualitatively to confirm that sensemaking in combination with ANSI/HFES-100 may help reduce HF/E in GCS.

Figure 13.6 shows the three phases in order.

FINDINGS

PHASE I – UAS GCS IO DEVICES

For quantitative phase I, 45 UAS-IO-1S surveys were distributed to program managers, engineers, and operators to identify UAS GCS IO device used in Groups 2 through 5. Figure 13.7 shows the phase I process flow chart:

Responses from 36 UAS GCS operators were collected and their responses were confirmed. The remaining 9 UAS-IO-1S surveys were inadequately completed or follow-up UAS-IO-1Q questioner was not completed/returned. The 36 participants represented as follows: 10 UAS GCS from Group 2, 7 from Group 3, 11 from Group 4, and 8 from Group 5. Figure 13.8 shows the UAS from each group that are part of this study:

The maximum flying experience for operators responding to UAS-IO-1S survey was 15 years while average was 8.5 years. Figure 13.9 shows the frequency of IO devices found in UAS GCS. "CW IO" are devices that were found in GCS and were covered by ANSI/HFES-100 standard's IO category, whereas "Non-CW IO" refers to devices not covered by ANSI/HFES-100.

The UAS-IO-1S surveys identified six IO devices (Figure 13.9, CW IO "blue") that were covered by ANSI/HFES-100. Gamepad was identified as an additional device; under write-in section of UAS-IO-1S (Figure 13.9, Non-CW IO "red"), gamepad was not covered by the standard. An analysis of each of the UAS GCS IO devices follows:

- Display: Device used in CWs to display (monitor) text, graphics, and data. Operators use display during takeoff, landing, and in-flight operations. It is a fundamental part of the GCS and operators rely on it for all UAS health, status monitoring, geo-location, and maneuvering. Display was found in all GCS, similar to the findings in previous study.

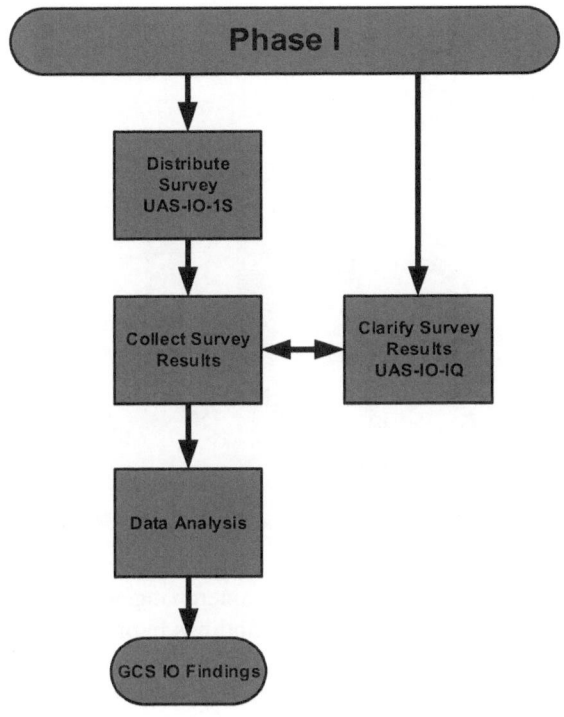

FIGURE 13.7   Phase I process flow chart.

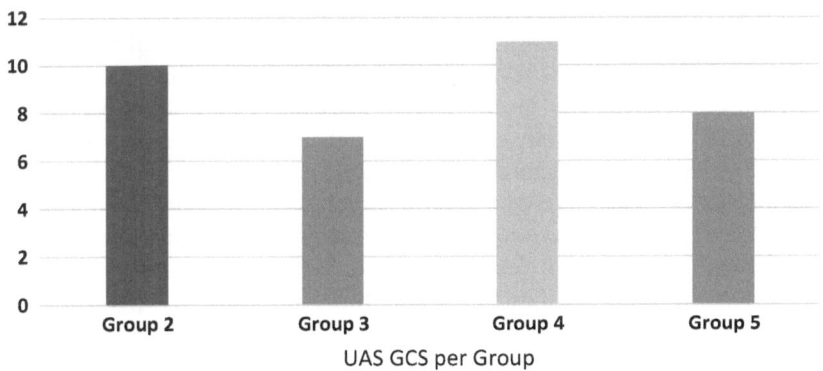

FIGURE 13.8   UAS in this study.

- Keyboard: Provides CW users with an interface to input alphanumeric data in computer. Operators use keyboard to enter flight configurations, from takeoff to landing. Keyboards were found in all GCS in this study similar to the findings in previous study.
- Mouse and trackball: Allows user to input data in CW by pointing, clicking, and other fine-grained adjustments. They are used by operators during an

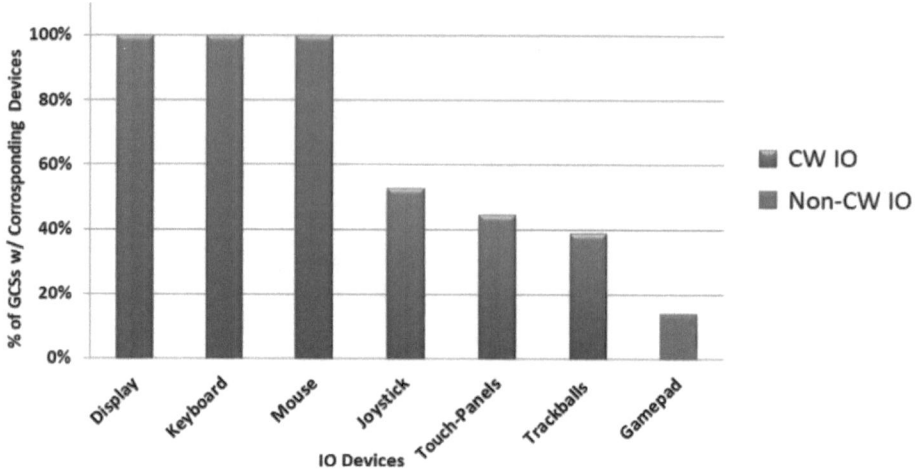

FIGURE 13.9   ANSI/HFES-100 compliant IO devices used in UAS GCS.

entire flight. Mouse and trackballs are interchangeable. All GCS studied had a mouse or trackball similar to the findings in previous study.

- Joystick: A hand-operated stick that pivots on a base, reports its angle and/or direction to the CW. A joystick is similar to center stick used by pilots in the aircraft. Operators use joystick to input angle and direction into the GCS; it is used for in-flight maneuvers only. Joysticks were found in 53% of the GCS in the study, which is a nominal increase from 50% in previous study (Waraich, Mazzuchi, Shahram, & Rico, 2013).
- Touch panel: A touch-sensitive panel for controlling a system without the use of any input devices (i.e., keyboard, mouse, or trackball). Touch panels are used during the entire flight. Touch panels were found in 44% of the UAS GCS in this study compared to only 25% usage in previous study (Waraich, Mazzuchi, Shahram, & Rico, 2013).
- Gamepad: Gamepads are supplied with gaming consoles, such as PlayStation and Xbox. In gaming consoles, a gamepad provides means to control an on-screen object or allow users to move through menus to make selections. Its use in the GCS is limited to surfing through menus and making selections. Gamepads are used during the entire flight. Gamepads were found in 14% of the GCS, which is a nominal increase from 10% in previous study (Waraich, Mazzuchi, Shahram, & Rico, 2013).

Table 13.2 shows IO devices used in the GCS from each of the UAS Groups 2 through 5.

Of the 36 GCS studied, there were a total of 225 IO devices or on average 6.25 IO devices per GCS. Five gamepads were found to be noncompliant to ANSI/HFES-100 IO category. Overall, 98% of all IO devices used in the UAS GCS are specified by ANSI/HFES-100 IO category. The previous study of 20 GCS had combined 127 IO devices, or 6.35 IO devices per GCS on average, which is slightly more than the 6.25 IO devices per GCS in this study. The slight downtick in use of number of IO devices

TABLE 13.2

IO Device Usage in Groups 2 through 5 UAS GCS

IO Devices	Group 2 (%)	Group 3 (%)	Group 4 (%)	Group 5 (%)	Overall (%)
Display	100	100	100	100	100
Keyboard	100	100	100	100	100
Mouse	100	100	100	100	100
Joystick	70	86	36	25	53
Touch panel	10	71	55	50	44
Trackball	40	43	27	50	39
Tablet and overlay	0	0	0	0	0
Puck device	0	0	0	0	0
Stylus/light pen	0	0	0	0	0
Gamepad	30	0	9	13	14

per GCS may have resulted from increased use of touch panels. Thus, resulting in elimination of some point and click hardware. Therefore, it is safe to conclude that the new study data do not have any significant change from the previous study. Thus, it proves ANSI/HFES-100's IO category may still be applicable to all UAS GCS IO applications (Waraich, Mazzuchi, Shahram, & Rico, 2013). Based on this information, we can accept Hypothesis 3 that ANSI/HFES-100 IO specifications apply to UAS GCS IO devices.

The IO devices employed by Groups 2–5 of UAS GCS are similar to the ones listed in the CW "ANSI/HFES-100" standard's IO category.

Phase II – GCS & CW IO Device Usage

For quantitative phase II, 48 UAS-N/E-2S surveys were distributed to UAS operators to assist in evaluation of IO device's use during "Normal Operation" and "Emergency Operation" in the UAS GCS. The operators were asked to evaluate each IO device against physical shape, functionality, physical settings, and software settings. Figure 13.10 shows the phase II process flow chart:

Responses from 41 UAS GCS operators were collected; 7 surveys were not returned. Table 13.3 shows response data for normal operation of ANSI compliant IO devices.

For the display, keyboard, mouse, trackball, and touch panel, there was a consensus among all operators that the UAS GCS IO device's usability under normal operating conditions is exactly the same as their usability in CWs. The joystick had the most variation for physical shape, functionality, physical settings, and software settings. Overall, the joystick in UAS GCS under normal operation is 90% similar, if not identical to its use in the CW.

Table 13.4 shows response data for emergency operation of ANSI-compliant IO devices. In the survey, operators were asked to recall the use of each type of device for an unexpected UAS GCS emergency situation. Most GCS operators had experienced

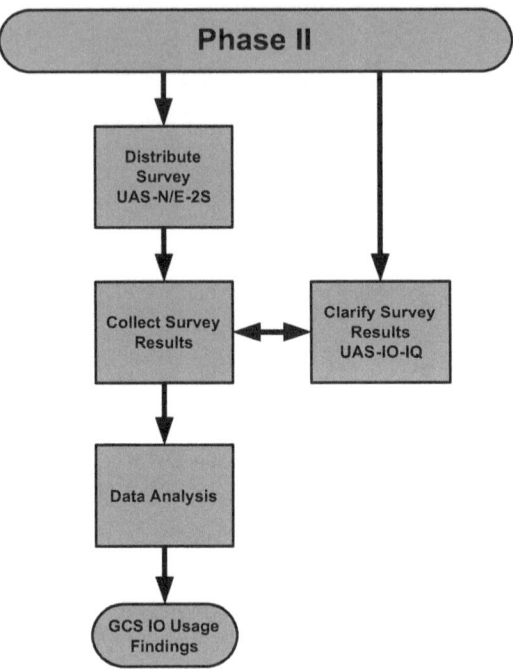

FIGURE 13.10   Phase II process flow chart.

TABLE 13.3

UAS GCS-CW IO Comparison (Normal Operation)

IO Devices	Variables				Normal Operation Device Similarity (%)
	Physical Shape (%)	Functionality (%)	Physical Settings (%)	Software Settings (%)	
Display	100	100	100	100	100.00
Keyboard	100	100	100	100	100.00
Mouse	100	100	100	100	100.00
Joystick	91	79	92	98	90.00
Touch panel	100	100	100	100	100.00
Trackball	100	100	100	100	100.00

an emergency or a UAS mishaps at some point in their recent past. Therefore, they evaluated the suitability of each of the IO devices in emergency situations.

The results remained unchanged for display, keyboard, mouse, trackballs, and touch panel between normal and emergency operation. The joystick functionality scored the lowest while the rest remained the same. A questioner UAS-N/E-2Q was used to clarify operators' response on joystick functionality. Nearly all operators

TABLE 13.4

UAS GCS-CW IO Comparison (Emergency Operations)

	Variables				
IO Devices	Physical Shape (%)	Functionality (%)	Physical Settings (%)	Software Settings (%)	Emergency Operation Device Similarity (%)
Display	100	100	100	100	100.00
Keyboard	100	100	100	100	100.00
Mouse	100	100	100	100	100.00
Joystick	91	73	92	98	88.50
Touch panel	100	100	100	100	100.00
Trackball	100	100	100	100	100.00

TABLE 13.5

UAS GCS-CW Comparability Mann-Whitney U Test Results

Hypothesis Test	Results
Sample size	41
Alpha, α	0.05
P-value	0.843
Decision	**Accept H$_2$**

hinted at the significance of accurate input since the consequences of inputting wrong information could result in a multimillion-dollar mishap.

The data from UAS GCS' normal and emergency operations analysis are equal in sample size and were independent. A Mann–Whitney U test with an α of 0.05 was used to test the hypothesis (see Table 13.5).

The Mann–Whitney U test results showed acceptance of hypothesis H2. With an α of 0.05, there is 95% confidence interval in the test results. Based on this information, we can accept Hypothesis 3. That usability of six IO devices studied in phase II is similar in both UAS GCS and CW.

The "ANSI/HFES-100 2007" standard's IO category applies to Groups 2 through 5 UAS GCS IO interfaces.

PHASE III – SENSEMAKING & ANSI/HFES-100 APPLICATION

For qualitative phase III, a Freedom of Information Act request was submitted to obtain a redacted version of a Predator mishap report from 2016. The AIB report was based on DoD HFACS model. The report is typically redacted for information that is deemed sensitive (such as operator names, mishap location, time, and date). Therefore, the actual operator of the UAS at the time of the mishap cannot be contacted to clarify information contained in the report. Based on the needs of this study, the information pertaining to the HF/E and IO devices was extracted from the mishap report. For the accuracy of this study, it is imperative that someone with knowledge of the Predator UAS GCS be able to answer questions related to the usage/layout of IO devices listed in mishap report. Therefore, a volunteer Predator operator was enlisted. He agreed to review the IO-related mishap report data and answer questions via email or teleconference. Figure 13.11 describes the phase III process flow chart for the application of sensemaking data/frame model and ANSI/-HFES-100 standard to evaluate the HF/E portion of the AIB report:

Once the IO-related information from AIB report was transcribed into 20 segmented chunks of data (SCD), where each segment represented a single idea, the

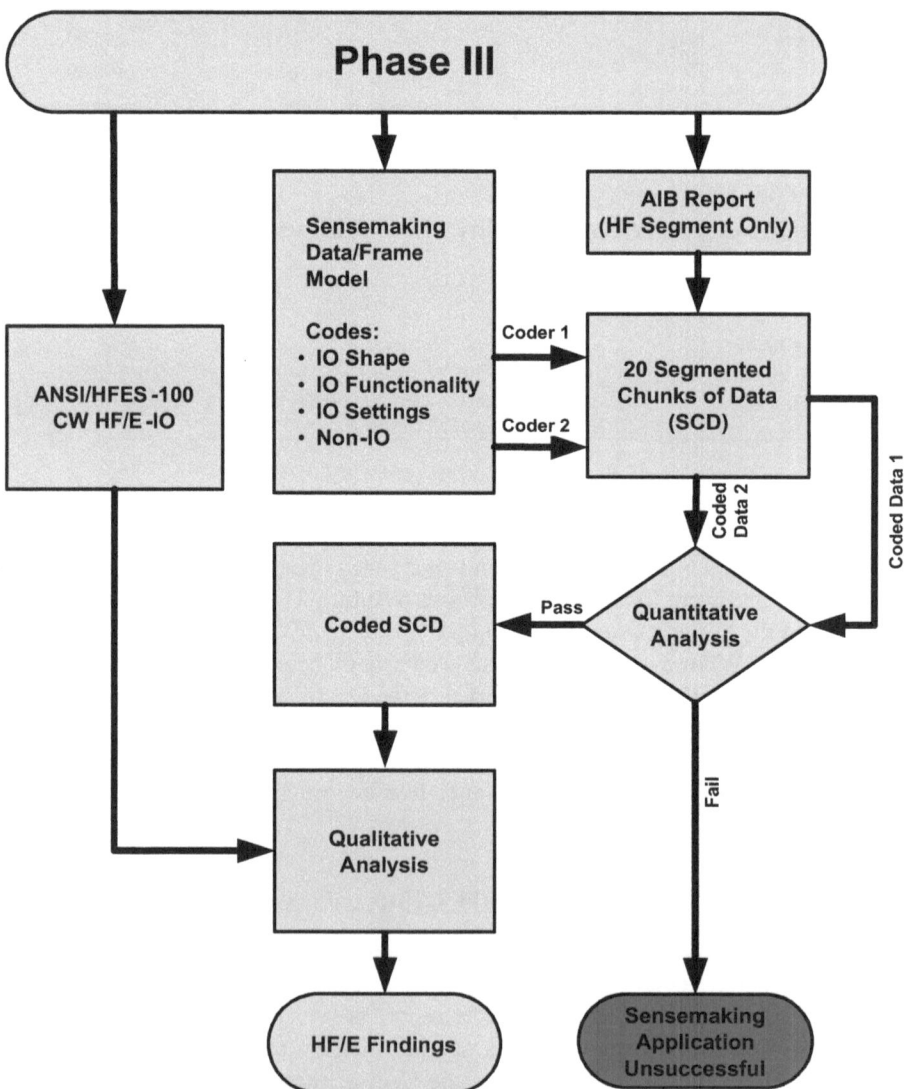

FIGURE 13.11 Phase III process flow chart.

order of all 20 SCDs was maintained to keep the context of the event. Two coders were selected to apply the sensemaking data/frame model to data chunks and to help establish the decision-making process of the original operator during mishap. Once each SCD was settled, both coders independently assigned it a final code. The two coders selected were the author (experienced in human factors application) and a retired 911 emergency dispatch operator. Both coders were knowledgeable about the sensemaking theory. Four codes listed in Table 13.6 were selected to gain insight into IO-related activities mentioned in the mishap report.

All 20 SCDs were coded independently by both coders. To reconcile any differences and to form a consensus agreement on all codes, a teleconference between both coders and the volunteer Predator operator was arranged. The SCDs from both coders were matched and disagreements on code assignment were discussed.

To access the coding reliability, the final results from both coders were compared by calculating Cohen's kappa and correlation coefficient. The results are in Table 13.7:

The Pr(a) represents a 90% of observed agreement between both coders, without any need for reconciliation. The remaining 10% had disagreements that were not reconciled. The Pr(e) of 45% represents a hypothetical by chance agreement between both coders. A Cohen's kappa of 0.61 and a correlation coefficient of 0.80 or above is considered as a substantial evidence of reliable rigorous coding process (Landis & Koch, 1977). For phase III of the study, kappa of 0.818 is well above 0.61, indicating an almost perfect agreement between the coders. The correlation coefficient of 0.773 is slightly lower than 0.80, but it is still considered to be a strong agreement between both coders.

Next, the SCDs were analyzed for further refinement and extraction of HF/E-related clues. From initial 20 SCDs, 12 were removed for being coded as "Non-IO"

TABLE 13.6
Sensemaking Data/Frame Model Codes

Codes	Definitions
IO shape	Shape/size of the IO devices caused confusion or limitation
IO functionality	Functionality/modes of the IO device caused confusion or limitation
IO settings	Physical adjustment of IO device caused confusion or limitation
Non-IO	Unrelated to the IO devices

TABLE 13.7
Coen's Kappa and Correlation Coefficient

Cohen's Kappa and Correlation Result	
Pr(a)	0.90
Pr(e)	0.45
K	0.818
R	0.773

and 2 were removed for disagreement in code assignment. From the remaining 6 SCDs, only 2 were of interest. Both coders had assigned an "IO Functionality" code to those SCDs.

The two SCDs could be summarized as a lost receive signal from the UAS, along with improper use of checklists as the main cause for the mishap. Due to the coders limited knowledge of the GCS layout, the Predator operator was asked to clarify the SCDs. The operator clarified that Predator GCS consists of two identical/interchangeable Pilot Payload Operator (PPO) stations, named PPO-1 and PPO-2. When the UAS is in air, one of the PPO controls the UAS while the other operates the camera when a PPO is configured in UAS control mode. Condition lever in forward position allows fuel to the engine. When in middle position, it shuts off engine's fuel supply (eventually shutting off the engine). When in aft position, it adjusts propeller to reduce drag. On the other hand, when the PPO is configured in payload control mode, the condition lever is used for camera iris settings, where, forward position opens the iris, middle locks the iris and back narrows the iris. See Table 13.8 for clarification:

During the mishap, PPO-1 was configured UAS control and PPO-2 for camera control. In PPO-2, the condition lever was set to the middle to lock camera iris. When the operator in PPO-1 tried to transfer the UAS over to the PPO-2, he did not properly follow the checklist before starting the procedure to switch from PPO-1 to PPO-2. In terms of sensemaking, the operator failed to confirm and elaborate his frame.

When the transfer from PPO-1 to PPO-2 occurred, the GCS lost the receive link from the UAS while the transmit link (from GCS to UAS) continued to work. Once the UAS transferred over to PPO-2, the condition lever in middle position sent the close fuel command over the working transmit link. Due the lost receive link, neither PPO-1 or PPO-2 updated the UAS status on display. The operator decided to reboot GCS (both PPO-1 and PPO-2) to allow the UAS to initiate a lost link recovery mode. Again, the operators failed to elaborate the fame and continued to use seat-of-the-pants approach. The command to close the fuel had already been sent to UAS, eventually shutting off its engine as it spiraled downward to its demise.

In this case, mishap report had described lost receive signal and improper use of a checklist as the cause of the mishap. Although, further investigation using sensemaking highlighted additional information that may not have been possible to extract when viewing just the mishap report, a single "condition-lever" was used for two modes of operation, which can be called Camera Mode and Flight Mode. If condition lever is in flight mode and then switched over to the camera mode, then it has no ramifications for the UAS flight. On the other hand, when same condition lever is left

---

TABLE 13.8

PPO Condition Lever Position Vs. Functionality

PPO	Condition Lever Forward	Condition Lever Middle	Condition Lever Aft
PPO-1 (UAS)	Open fuel	Close fuel	Adjust propeller to reduce drag
PPO-2 (Camera)	Open camera iris	Lock camera iris	Narrow camera iris

in middle position on camera mode and then switched over to flight mode, then the condition lever controls the UAS fuel and can shut down its engine midflight, leading to catastrophic results.

Based on this newfound information, one can argue that the AIB mishap report is accurate. Nevertheless, the condition lever mode information cannot be simply ignored. This significantly highlights the fact that such a safety-critical function of controlling the fuel flow to the engine shall not be tied to a condition lever which is being used for other nonflight essential purposes in another modes.

When trying to apply ANSI/HFES-100 as it currently stands, there is no direct reference to condition lever. Although the standard does state that when keys have collateral function, their mode of operation shall be clearly indicated (ANSI/HFES-100, 2007). Therefore, the combined application of sensemaking and ANSI/HFES-100 standard can not only capture the hidden HF/E in the GCS but may also find a viable solution for such issues.

That is, the application of sensemaking to highlight the hidden control lever mode-based HF/E shortfalls in the GCS design, coupled with the application of ANSI/HFES-100 IO category of the standard to remedy such shortfalls, thus, possibly leading to a reduction of HF/E in GCS and may reduce HF/E-associated UAS mishaps.

Therefore, with the validation of H3 and H2 in phases I and II, respectively, along with the findings in phase III, we can conclude that the H1 is also accepted.

The application of sensemaking coupled with "ANSI/HFES-100" standard to help identify and resolve HF/E in GCS IO interface design for Groups 2 through 5 UAS GCS could minimize HF/E impact on UAS operation.

CONCLUSION

This study was conducted in three phases:

The purpose of the first two phases (phase I and phase II) was to reaffirm the findings of a previous similar study of 20 UAS GCS by increasing the number of UAS GCS studied to 36. It appears that the general findings of this study did not significantly deviate from the previous study. Both studies found that ANSI/HFES-100 provides HF/E specifications when six of the IO devices were found to be utilized in (CW based) UAS GCS from Groups 2 to 5. These IO devices allow operators to remotely support the UAS flight mission from takeoff to landing. Since, there is no direct flight control or a cockpit on board the UAS, any errors made while using these relatively inexpensive IO devices could be amplified several thousand folds. Many of the UAS from Group 4 or 5 have a multimillion-dollar price tag, resulting in a significant monetary loss. Both studies confirmed that 98% of all UAS IO devices are covered by the ANSI/HFES-100 standard. The average number of IO devices use per GCS was slightly lower at 6.25 compared with 6.35 devices per UAS in previous study, which was explained by the increase in the use of touch panels in GCS to 44% compared to the 25% in previous study. When touch panels are used, the need of a point and click device (i.e., trackball) drops. Although, not all errors are avoidable, they can be minimized by simply designing the UAS GCS in accordance with ANSI/-HFES-100 standard.

The third phase of this study successfully applied sensemaking data/frame model and ANSI/HFES-100 standard to a HFACS model-based AIB report. It highlighted at least one additional HF/E issue in the GCS, which was not mentioned as an influencing factor in the mishap AIB report. The phase III results found that the dual use of the condition lever that was inadvertently left in an incorrect mode resulted in shutting off fuel supply and causing the mishap. It was found to be counter intuitive that the same condition lever was used to control a camera iris in one mode and a fuel shutoff switch in another. The HFACS model-based report had listed lost receive signal from the UAS and an improper use of checklist as the reason for the mishap. The position of the control lever was not listed as the main cause of the mishap. Moreover, the application of ANSI/HFES-100 standard resulted in specification for dual mode controls. The standard states when keys have collateral function, their mode of operation shall be clearly indicated, which was obviously not the case in this GCS, resulting in an overlooked HF/E that was not captured by the HFACS model.

Moreover, this study helped demonstrate the applicability of sensemaking data/ frame model to extract information from the AIB reports that may have been overlooked previously. It also highlighted the fact that the current taxonomies/models used to investigate human factors in manned aircrafts may not be sufficient to investigate HF/E in UAS GCS. Adding sensemaking and human factors experts to the UAS mishap investigation process, or even in the interpretation of an AIB report, can provide a profound understanding of the true nature cause of an accident. Once the HF/E deficiency is found it can then be fixed retroactively.

Developing a resilient system is a lifelong iterative process where repeated application of sensemaking process will result in an HF/E-resilient system design. The real purpose of sensemaking is to provide meaningful feedback to the system when it is time to alter its understanding of problem situations. Therefore, the adaptive and resilient sensemaking process requires ways of identifying irregularities and conditions that require a border change in system design (Hoffman & Hancock, 2017).

REFERENCES

2018 RPAS Yearbook. (2018, October). *RPAS: The Global Perspective*. Paris, France: Blyenburgh & Co.

Agarwal, S., Murphy, R., & Adams, J. (2014). Characteristics of indoor disaster environments for small UASs. *2014 IEEE International Symposium on Safety, Security, and Rescue Robotics (2014)* (pp. 1–6).

Andersen, H., Bove, T., Isaac, A., Kennedy, R., Kirwan, B., & Shorrock, T. (2002). *Short Report on Human Performance Models and Taxonomies of Human Error in ATM (HERA)*. Brussels: EUROCONTROL.

ANSI/HFES-100. (2007). *Human Factors Engineering of Computer Workstations*. Santa Monica, CA: Human Factors and Ergonomics Society.

Baur, J. (2007). *The NASA Dryden Flight Research Center Unmanned Aircraft System Service Capabilities*. Washington, DC: NASA.

Berni, J., Zarco-Tejada, P., Suarez, L., & Fereres, E. (2009). Thermal and Narrowband Multispectral Remote Sensing for Vegetation Monitoring from an Unmanned Aerial Vehicle. *IEEE Transactions on Geoscience and Remote Sensing, 47*(3), 722–738.

Brittain, J. (2009). Electrical Engineering Hall of Fame: Charles F. Kettering. *Proceedings of the IEEE, 97*(10), 1737–1739.

Chanda, M., DiPlacido, J., Dougherty, J., Egan, R., Kelly, J., Kingery, T., ... Others. (2010). *Proposed functional architecture and associated benefits analysis of a common ground control station for Unmanned Aircraft Systems.* Naval Postgraduate School Monterey CA Department of Systems Engineering.

Corlett, E. N., & McAtamney, L. (1988). Ergonomics in the Workplace. *Physiotherapy, 74*(9), 475–478.

Creswell, J. (2009). Research Design: Qualitative, Quantitative, and Mixed Methods Approaches. *Human Factors: The Journal of the Human Factors and Ergonomics Society, 52*(1), 17–27.

Dekker, S. (2001). The re-invention of human error. *Human Factors and Aerospace Safety, 1*(3), 247–265.

Dekker, S. (2004). *Ten Questions about Human Error: A New View of Human Factors and System Safety.* Boca Raton, FL: CRC Press.

Gawron, V. J. (1998). Human factors issues in the development, evaluation, and operation of uninhabited aerial vehicles. *Paper presented at the Proceedings of the Association for Unmanned Vehicle Systems International '98.*

Gioia, D. (2006). On Weick: An Appreciation. *Organization Studies, 27*(11), 1709–1721.

Gordon, P., Jeffries, J., & Flin, R. (2002). Designing a Human Factors Investigation Tool to Improve the Quality of Incident Investigations. *SPE International Conference on Health, Safety and Environment in Oil and Gas Exploration and Production.*

Guidance, I. (2008). *08-01, Unmanned Aircraft Systems Operations in the US National Airspace System.* Federal Aviation Administration.

Haines, P. (2007). Satellite Communications with UAVs: Practical Lessons from High Speed Trains and Planes. *Paper presented at the 2007 IET Seminar on Communicating with UAV's.*

Hamilton, S., & Stephenson, J. (2016). *Testing UAV (drone) aerial photography and photogrammetry for archeology.* Lakehead University.

Hanson, W., Creswell, J., Clark, V., Petska, K., & Creswell, D. (2005). Mixed Methods Research Designs in Counseling Psychology. *Journal of Counseling Psychology, 52*(2), 224–235.

Hobbs, A., & Shively, R. (2014). Human Factor Challenges of Remotely Piloted Aircraft. *31st EAAP Conference* (pp. 5–14). Valletta, Malta.

Hoffman, R., & Hancock, P. (2017). Measuring Resilience. *Human Factors, 59*(4), 564–581.

Kaste, K. (2012). *Naturalistic Study Examining the Data/Frame Model of Sensemaking by Assessing Experts in Complex, Time-Pressured Aviation Domains.* Embry-Riddle Aeronautical University - Daytona Beach.

Kilskar, S. S., Danielsen, B.-E., & Johnsen, S. O. (2020). Sensemaking in Critical Situations and in Relation to Resilience—A Review. *ASCE-ASME Journal of Risk and Uncertainty in Engineering Systems, Part B: Mechanical Engineering*, 1–10.

Klein, G., Phillips, J., Rall, E., & Peluso, D. (2007). A Data–Frame Theory of Sensemaking. *Expertise Out of Context: Proceedings of the Sixth International Conference on Naturalistic Decision Making* (pp. 118-160). Philadelphia, PA: Taylor & Francis Inc.

Landis, R. J., & Koch, G. G. (1977). The Measurement of Observer Agreement for Categorical Data. *Biometrics*, 159–174.

Maitlis, S., & Christianson, M. (2014). Sensemaking in Organizations: Taking Stock and Moving Forward. *Academy of Management Annals, 8*(1), 57–125.

Manning, S. D., Rash, C. E., LeDuc, P. A., Noback, R. K., & McKeon, J. (2004). *The Role of Human Causal Factors in U.S. Army Unmanned Aerial Vehicle Accidents.* Ft. Rucker, AL: U.S. Army Aeromedical Research Laboratory.

Matolak, D., & Sun, R. (2015). Unmanned Aircraft Systems: Air-Ground Channel Characterization for Future Applications. *IEEE Vehicular Technology Magazine, 10*(-2), 79–85.

Miessner, B. (1916). *Radiodymanics: The Wireless Control of Torpedoes and other Mechanisms*. New York: D. Van Nostrand Company.

Mouloua, M., Gilson, R., Kring, J., & Hancock, P. (2001). Workload, Situation Awareness, and Teaming Issues for UAVUCAV Operations. *Human Factors and Ergonomics Society Annual Meeting Proceedings, 45*, 162–165.

Nas, M. (2008). *The Changing Face of the Interface: An Overview of UAS Control Issues & Controller Certification, Unmanned Aircraft Technology Applications Research (UATAR) Working Group*. 27: Murdoch University.

Nisser, T., & Westin, C. (2006). *Human Factors Challenges in Unmanned Aerial Vehicles (UAVs): A Literature Review*. Lund, Sweden: Lund University.

Pastor, E., Lopez, J., & Royo, P. (2007). UAV Payload and Mission Control Hardware/Software Architecture. *Aerospace and Electronic Systems Magazine, IEEE, 22*(6), 3–8.

Peter, N., & Karl, G. E. (2016). Identifying and Mitigating Human Factors Errors in Unmanned Aircraft Systems. *16th AIAA Aviation Technology, Integration, and Operations Conference*, Washington, DC.

Rogers, B., Palmer, B., Chitwood, J., & Hover, G. (2004). *Human-Systems Issues in UAV Design and Operation*. Wright-Patterson AFB, OH: Human Systems Information Analysis Center.

Sauter, S., Schleifer, L., & Knutson, S. (1991). Work Posture, Workstation Design, and Musculoskeletal Discomfort in a VDT Data Entry Task. *Human Factors, 33*(2), 151–167.

Sheridan, T. (1992). *Telerobotics, Automation, and Human Supervisory Control*. Cambridge, MA: MIT Press.

Sieck, W., Klein, G., Peluso, D., Smith, J., & Harris-Thompson, D. (2007). *FOCUS: A Model of Sensemaking*. Arlington VA: U.S. Army Research Institute for the Behavioral & Social Sciences.

Tesla, N. (1898). *USA Patent No.: 613.809*.

Thompson, W. T., & Tvaryanas, A. P. (2008). *Unmanned Aircraft System (UAS) Operator Error Mishaps: An Evidence-based Prioritization of Human Factors Issues*. Brooks City, TX: North American Treaty Organization (NATO).

Thompson, W., & Tvaryanas, A. (2005). *U.S. Military Unmanned Aerial Vehicle Mishaps: Assessment of the Role of Human Factors Using HFACS*. Brooks City-Base, TX: Air Force Research Laboratory.

Wagenaar, W., & Groeneweg, J. (1987). Accidents at Sea: Multiple Causes and Impossible Consequences. *International Journal of Man-Machine Studies, 27*(56), 587–598.

Waraich, Q., Mazzuchi, T. A., Shahram, S., & Rico, D. F. (2013). Minimizing Human Factors Mishaps in Unmanned Aircraft Systems. *Ergonomics in Design*, 25–32.

Weatherington, D. (2010). Unmanned Aircraft Systems. *Paper presented at the Precision Strike Annual Review*.

Weick, K., Sutcliffe, K., & Obstfeld, D. (2005). Organizing and the Process of Sensemaking. *Organization Science, 16*(4), 409–421.

Wiegmann, D., & Shappell, S. (2001). Human Error Perspectives in Aviation. *International Journal of Aviation Psychology, 11*(4), 341–357.

Williams, K. (2004). *A Summary of Unmanned Aircraft Accident/Incident Data: Human Factors Implications*. Oklahoma City, OK: Civil Aerospace Medical Institute, Federal Aviation Administration.

Zhang, W., Feltner, D., Shirley, J., Kaber, D., & Neubert, M. (2020). Enhancement and Application of a UAV Control Interface Evaluation Technique. *ACM Transactions on Human-Robot Interaction, 9*(2), 1–20.

14 Constrained Autonomy for a Better Human– Automation Interface

Ø. J. Rødseth
SINTEF Ocean

CONTENTS

SHIP AUTONOMY AND HUMAN CONTROL

The concept of autonomous or uncrewed ships is not new. Japan investigated remote control of ships in the "Highly reliable intelligent ship" project from 1982 to 1988 (Hasegawa 2004). The rocket launching platform *L/P Odyssey*, classified as a mobile offshore unit (MOU), was remotely controlled during the launch phase. Thus, it operated as a de facto uncrewed ship in international waters from 1999 to 2014 (Tass 2018). The first large-scale study on uncrewed and autonomous merchant ships was the EU project MUNIN, running from 2012 to 2015 (Rødseth & Burmeister 2012). Since then, there has been a steady increase in new investigations and concept studies. *M/S Yara Birkeland* is probably the best known and is at the time of writing planned to operate autonomously and uncrewed from 2022 (Yara 2018). A major benefit of ship autonomy is that the ship can be uncrewed, although uncrewed operation can also be achieved through remote control as for *L/P Odyssey*. Uncrewed ships save capital cost when removing the living quarters and life support systems from the ship; it can save crew cost and it allows new and innovative designs of the ship (Rødseth 2018).

The word autonomy comes from the Greek roots *autos*, "self," and *nomos*, "law," and literally means the freedom to make one's own laws. For an autonomous mobile robotic system like an autonomous ship, there are several suggested definitions of autonomous and the subject will be discussed later in this chapter. An autonomous ship can be classified as an industrial autonomous system. This is an autonomous unit, or a collection of such, that can operate safely and efficiently in a real-world environment while doing operations of direct commercial value and which can be manufactured, maintained, deployed, operated and retrieved at an acceptable cost relative to the value it provides (Grøtli et al. 2015). When operating in general seaways together with other ships and leisure crafts, this puts a high demand on safety and reliability that is difficult to achieve with automation systems today. Furthermore, merchant ships have a high capital value, and it is expected that most autonomous and uncrewed ships will be continuously supervised from a remote control centre (RCC) to keep a close watch on the ship and the corresponding investment. However, when an RCC is in place, it also makes sense to let the RCC operators participate in the control of the ship. This avoids the need for the automation system to be able to handle all possible operational cases as the operator is available for the cases that are too complex for the automation to handle reliably.

This means that most autonomous ship systems will involve both an automation system and a human operator. Thus, the question of a how to design a high-quality human–automation interface (HAI) is a central one for autonomous ships. This chapter will discuss some possibilities for the design of the automation system for autonomous ships that may enable a better HAI to be designed. In the following, the term autonomous ships will be used for an automated ship where human operators are available but are not continuously attending to the control positions. The operators may be on the ship or in the RCC.

CONSTRAINED AUTONOMY IN THE LITERATURE

The form of constrained autonomy discussed in this text was first published as a concept in (Rødseth & Nordahl 2017). Here, it described a designed-in limitation on the possible action of an automation system in a mixed autonomy/human operator context. The objective is to create a more deterministic behaviour as seen from the human designer or operator, and by that make the allocation of tasks and responsibilities between human and automation more efficient and safer.

Other writers have used a similar terminology for other concepts such as in Al-Rifaie et al. (2012) where it applies to Gaussian constrained autonomy in swarms, where constrained refers to a limited random behaviour by swarm members. In Jha et al. (2018), the term chance-constrained temporal logic is used on a variant of temporal logic adapted to perception uncertainty. Both these uses of constrained autonomy are very different from the concept as it is described here.

The terms limited autonomy and partly autonomous have also been used frequently in the literature, but this normally refers to emergent and generally unwanted limitations in the automation system and not to a design feature.

AUTOMATION, AUTONOMY, RESPONSE TIME AND DEADLINE

Automation and autonomy has a wide range of definitions in the literature (see, e.g., Vagia et al. 2016). For the purposes of this chapter, a relatively simple definition will be used. Here, automation can be defined as "pertaining to a process or device that, under specified conditions, can function without human intervention". Furthermore, this can be used to define autonomy as "in the context of ships, autonomy e.g. as in 'Autonomous Ship', means that the ship uses automation to operate without human intervention, related to one or more ship processes, for the full duration or in limited periods of the ship's operations or voyage." These are definitions that have been proposed to the International Maritime Organization (IMO) by ISO (2020a). These definitions simply say that automation can be used to provide autonomy and that autonomy emerges when the system is *designed, approved* and *deployed* to be operated without human intervention or supervision for certain periods. This could also be used to describe levels of autonomy by how long the operator can stay away as indicated below.

For a relatively simple automation system that is able to detect the danger of a collision, but not to make reliable corrective actions, this could be illustrated along a time axis as in Figure 14.1. Here, the danger of collision is measured in Closest Point of Approach (CPA – typically measured in nautical miles, nm) and Time to CPA (TCPA).

When a danger of collision is detected, the automation system needs to alert the crew to this so that they can take evasive actions. The crew have been organized to arrive and be ready at the control position, at the latest at T_{MR}, which is defined as the crew's *maximum response time*. The deadline for the crew's response is given by the actual situation and is defined as the *response deadline*, T_{DL}. A minimum requirement for safe operation is that T_{DL} is longer than T_{MR}. Some examples of different crew organizations are given below, where response times are only indicative and given as examples:

- *Operator in control:* The operator is directly in control of the ship. Hand-over time is not relevant ($T_{MR}=0$).
- *Operator supervision:* Automation is used to assist operator, and operator is overseeing the operation and needs only a short time to gain situational awareness when actions are needed ($T_{MR}\approx 10$s of seconds).
- *Operator at site*: An operator is at the control position but is working with other tasks and will need time to gain situational awareness. This could be on the order of a minute or so ($T_{MR}\approx$ minutes).
- *Operator available:* The operator is available, but is in another location, possibly sleeping, and will need several tens of minutes to reach the control position and to regain safe control ($T_{MR}\approx 10$s of minutes).
- *No operator:* There is no operator and automation must be able to handle all operations by itself (T_{MR} is the duration of the operation or the voyage).

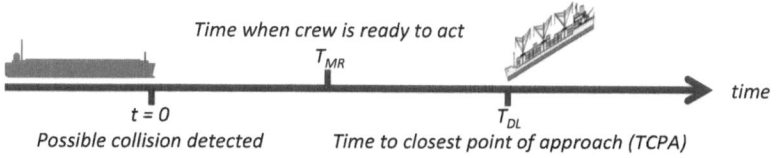

FIGURE 14.1 Simple autonomy with crew support.

Note that this form of characterization of response time is independent of the crew residing on board or in an RCC, although the times will likely be different in the two situations due to differences in equipment and access to detailed sensor or situation information.

Today, most ships have an autopilot or a track pilot. Open sea with no ship or other objects in the vicinity allows the officer of the watch to be away from the bridge for relatively long periods. With the above definition of automation and autonomy, this makes a ship controlled by an autopilot autonomous with respect to the process of keeping a steady speed and course in open sea. Autonomy in this context may seem counter-intuitive but is related to the low abstraction level on the involved functions. In the work presented here, four levels of functional abstraction are used:

1. **System objectives**: This is the highest abstraction level and is associated with generating the objectives for the design of the autonomous ship system. This may in some cases be static or at least have a long horizon, e.g. transport available cargo between ports A and B. This will be the basis for the design of the control system.
2. **Planning:** This is also a high abstraction level and will normally be an external input to the ship control system. It is related to the overall planning of the voyage or mission within the constraints of the system objectives. In most cases, this is expected to be supplied by the ship operators.
3. **Goal based**: These can be seen as a sequence of process goals for the autonomous ship system. Each goal is expected to be associated with one or more processes or tasks. This is also the most likely abstraction level for commands to the autonomous ship system.
4. **Functional**: This is specific instructions to a function, such as an autopilot. This is normally on a low abstraction level and will not normally be used as commands to the autonomous ship system.

As exemplified above, autonomy on the functional level already has been developed and is used in well-controlled environments such as autopilots on high sea or car cruise controls on highways. The goal of the work presented here is to contribute methods to extend autonomy to higher abstraction levels while giving human operators a better understanding of the capabilities and limitations of the automation system.

SENSEMAKING AND TRUST IN AUTOMATION

One basic issue in the HAI is its ability to support a proper level of operator's trust in the automation system (Lee & See 2004). This should not be too low, leading to disuse of the automated functions and neither should it be too high, leading to over-reliance and misuse of the automation. In addition, the operator must be able to make sense of the relationship between automation, his or her responsibilities and the situation at hand. The latter could be called "sensemaking", which can be defined as "a motivated, continuous effort to understand connections (which can be among people, places, and events) in order to anticipate their trajectories and act effectively" (Klein et al. 2006).

This chapter will not try to link the concept of constrained autonomy to human–machine interface (HMI) research as that is clearly outside the author's expertise. However, the proposal put forward here is that both trust and sensemaking to some degree may be linked to the relationship between T_{MR} and T_{DL}. If the system consistently is able to determine T_{DL} and alert the operator before T_{MR} elapses, one can argue that the operator should get more consistent trust in the automation system's ability, both to control the process under normal conditions and to warn the operator when something requires the operator's attention. It will be left to experts in the HMI field to validate this proposal and investigate what consequences these ideas have for the operator and for the design of the HMI. This issue is complex and it is difficult, if not impossible to draw any clear conclusions on what is the best strategy for building a suitable level of operator trust (Hoff & Bashir 2015). However, some issues that seem to give positive effects are determinism in automation responses, minimizing false alerts and making it as clear as possible what the automation system is able to do, what it actually does and where the operator's intervention is required. Again, it can be argued that a higher emphasis on the deadlines and response times may be important to achieve these objectives.

The remaining part of the chapter will concentrate on the technical aspects of constrained autonomy and how it can be implemented.

THE OPERATIONAL ENVELOPE

The operational envelope (OE) can be defined as "The specific conditions under which a given autonomous ship system is designed to function, including, but not limited to, its environmental conditions and the different mission or voyage phases, as well as all anticipated failures." The definition of OE is based on the concept of the "Operational Design Domain" that was defined in SAE J3016 (2016) and developed further for use on autonomous ships in Rødseth (2018). The name has later been changed to operational envelope (OE) during the work on a standard terminology for autonomous ships (ISO 2020b).

The OE will be directly linked to the Ship Control Tasks (SCT) which will specify the details of the different tasks or processes to be performed. The OE and SCT will also specify the division of responsibilities between humans and automation. The OE and SCT can be defined on basis of a "concept of operations" document, or CONOPS. Figure 14.2 is a simplified object diagram that illustrates objects and relationships related to the OE and SCT.

The two large boxes at the bottom left are the constraints on the OE given by operational limitations in the ship system as well as the properties of the environment. Additional constraints will be added by the concept of operation, e.g. the ship cannot operate at night or during wintertime (phases and functions). The same factors that define the constraints will also play a role in determining the dynamic conditions for SCT.

The darker boxes represent the OE itself as well as the additional fall-back space which contains minimum risk conditions (MRCs). MRCs will be activated when the limits of the OE are exceeded. The OE will normally be divided into subdivisions, usually based on the mission phases and the relevant ship processes. As an example,

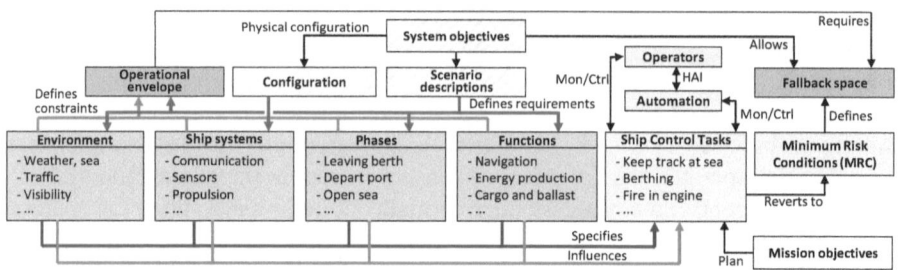

FIGURE 14.2 A simplified ER-diagram showing relationships to operational envelope.

there will normally be different subdivisions for navigation in open sea and navigation in port areas. In addition, non-controllable constraints, such as weather or anticipated technical problems may also require further subdivisions in the OE. Different MRCs can be associated to each of these subdivisions.

The lighter parts of the diagram to the right represent the functional part of the ship system. This will be realized through the SCT. SCT may be executed by operators or automation, and if both are involved, there also needs to be a Human–Automation Interface (HAI) between them. The crew and automation will execute the SCT based on the mission or voyage plan, taking ambient conditions into consideration. For automated operations, it may also be necessary for the operator to specify dynamic constraints for SCT e.g. do not exceed 12 knots or do not allow a cross-track deviation from planned route by more than one nautical mile.

THE OPERATIONAL ENVELOPE AS A STATE SPACE

The OE and the SCT exist in a multidimensional state space that will be called S in the following. Any condition that the ship can end up in is a state vector c in S. As has been indicated and as will be discussed later, OE will normally be discretized into a finite number of smaller subdivisions or sub-spaces as illustrated in Figure 14.3. In the following, OE will be denoted as O and a sub-space in O will be denoted as O_n. It is important that O covers exactly the same state space as S, i.e. it can be said to be congruent with S. This is necessary to ensure that any condition c the ship can be in can be mapped to an appropriate number of OE states, so that any individual state variables in c map to one O_n. These relationships are presented formally in Eq. (14.1).

$$O \cong S$$

$$c = [c_1, c_2, \ldots, c_n]^T \tag{14.1}$$

$$\forall c \in S, \ \forall ci \in c, \ \exists O_n \subset O : c_i \in On$$

The active part of O will normally vary over time, as not all states are relevant in all mission phases or for all ship processes. Thus, O may have to be subdivided into separate components to reflect voyage phases (L: leaving berth, D: depart port; C: coastal, etc.) and different processes (V: voyage planning; S: sailing; O:

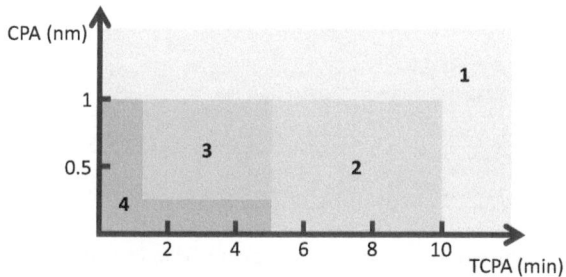

FIGURE 14.3 A discrete number of states over continuous state variables.

observation; F: Fire, etc.). This is shown in Eq. (14.2) where the different subdivisions of O have been given process as subscript and phase as superscript. The actual number of subdivisions will depend on the case at hand. Other principles for subdivisions can also be used. Each of the component spaces may have different number of dimensions.

$$O = O^L \cup O^C \cup \cdots \cup O^X$$

$$O^X = O^x{}_V \cup O^x{}_S \cup O^x{}_O \cup O^x{}_F \cup \cdots \cup O^x{}_Y$$

(14.2)

The dimensions of each O_n subdivision are defined by a number of continuous state variables such as CPA and TCPA (see Figure 14.1). O_n can be seen as a state space consisting of a number of possibly multi-dimensional states s, where each s is defined over a range of one or more state variables. This is illustrated in Figure 14.3, where four states are suggested for various combinations of the two state variables TCPA and CPA. These states are also the same as the states and corresponding variable ranges shown in Figure 14.4.

For a state s, it may be possible to determine T_{DL} for a given environmental and ship condition c. As c can vary while s is active, the value of T_{DL} will generally also vary inside s.

It may not always be possible to define T_{DL}, e.g. when various forms of artificial intelligence (AI) technologies are used, where one cannot say a priori that the task always will find a useable solution to a problem or in what time frame the solution will be found.

Figure 14.4 shows an example of a simplified state transition diagram for part of the sailing process, corresponding to the states in Figure 14.3 and the illustration in Figure 14.1. In this example, the state space vector consists of the variables TCPA and CPA. The figure also shows the value ranges of T_{DL} for each state.

States 1 and 2 allow autonomous operation if the crew's maximum response time T_{MR} is 10 minutes. States 3 and 4 will require operator assistance before TCPA goes below 1 minute, otherwise a fall-back state F1 will be activated, e.g. ordering the ship to stay still in the water. To allow the crew time to reach the bridge, state 3 must be defined so that it will have a T_{DL} of 10 minutes at the time state 3 is entered. In this case, one would have to alert the crew at the latest in the transition between states 2 and 3. State 4 is a state where the automation leaves the control responsibility to the

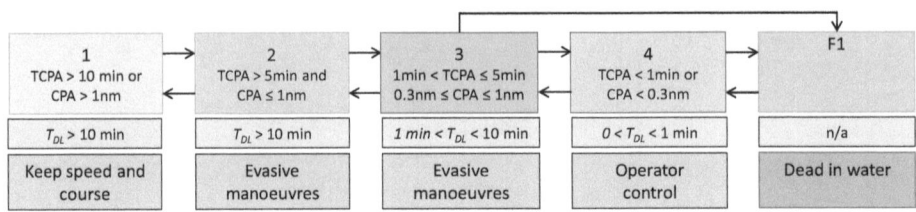

FIGURE 14.4 Examples of operational envelope (OE) states and ship control tasks.

human operator. Also, here the fall-back can be activated, e.g. if the operator does not respond.

The figure also shows how to establish a one to one relationship between the OE states and the SCT. The task associated with the fall-back state F1 is called a MRC as F1 is not part of the OE. The tasks and MRC are illustrated by the boxes at the bottom, below the state boxes. Note that the tasks corresponding to states 2 and 3 is identical in description but are still considered two different tasks as operational parameters are different.

One can also create a corresponding set of states for daytime sailing when crew is active and T_{MR} normally is shorter, e.g. 1 minute. One could keep the same pattern but reduce CPA and TCPA correspondingly. To keep the amount of code and specifications lower, this could be implemented as parameterized states and SCTs. This could be done by comparing TCPA to T_{MR} and distances to a relationship between speed and T_{MR}.

The above discussion shows that it is possible to split O into different partitions, where each partition can define different relationship between automatic and crew control. Note that this partitioning is not the same as the partitioning defined in Eq. (14.2). The different relationships between automation and crew are illustrated in Figure 14.5, with the following O components:

1. O_{FA} – Fully Autonomous: States that the automation system is designed to handle alone, without human interaction. T_{DL} is not relevant as long as the system remains in these states.
2. O_{AC} – Autonomous Control: States where the automation system can sustain automatic operation for at least a known T_{DL} without human support.
3. O_{OA} – Operator and Automation: States where automation can handle some situations, but where T_{DL} cannot be defined. A human needs to supervise automation and be ready to take over control.
4. O_{OE} – Operator Exclusive: States where direct operator control is required. The automation may assist the operator but is not generally able to control the ship in a safe manner.
5. F: Fall-back states containing the MRCs that are used in cases where events take the system out of O. This may happen due to unanticipated failures, environmental conditions outside O or failure of an operator to respond when the system requires human intervention.

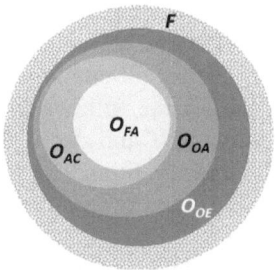

FIGURE 14.5   Operational envelope classification.

The partitions are drawn enclosed in each other to illustrate that restrictions on human presence gradually becomes more stringent as one enters the next level outwards. Fall-back states F are drawn outside O as they are not considered part of O.

THE PLAN STATE SPACE AND THE SYSTEM RESPONSE DEADLINE

O is a static description of the capabilities of the autonomous ship system. The mission objectives or the plan (see Figure 14.2) together with environmental conditions will specify what parts of this overall envelope will be used in a particular mission. O has previously been specified as a state space and a corresponding plan state space (P) can be defined. This will be a subset of S. The corresponding subset of O, here denoted $\hat{O}$ with subdivisions $\hat{O}_n$, may restrict the states and SCT that are used during the execution of the plan. The plan can be looked at as a discrete function of time $p(t)$ that returns a state vector from P corresponding to the condition the ship should have at the specified time step. These relationships are shown in Eq. (3).

$$P \subseteq S$$

$$\hat{O} \cong P \tag{14.3}$$

$$p(t) \in P$$

During the voyage or mission, P will restrict O and by that the "freedom" of the control system and crew, and this could also influence on the T_{DL} calculated from $\hat{O}$. One example is that additional geographic limitations such as a restricted fairway limits the ship's ability to avoid obstacles, and this would obviously have an effect on T_{DL} in state 2 of the system described in the previous section.

If necessary, the automation system needs to cater for this by recalculating T_{DL} for any new constraints that can apply. However, it should in many cases be possible to define subdivisions of O that has states which can be made independent of P restrictions. Typically, this would be used to factor out, e.g. geography-independent subdivisions, as geography is likely to be part of P restrictions. Then, the updated T_{DL} could be linked to the remaining geography related states.

At any point in time, an autonomous ship will have a number of active SCT with different degree of automation and different timing requirements. Examples are SCT related to navigation, energy production, and stability. This means that the operator's response deadline for the full system will be the minimum of the individual T_{DL} values for all active SCT and states. If a function FDL is defined to determine T_{DL} for a given state and condition, the system-wide T_{DL} can be defined as in Eq. (14.4).

$$\forall\, c_j \in c, \exists \hat{\boldsymbol{O}}_n \subset \hat{\boldsymbol{O}} : c_j \in \hat{\boldsymbol{O}}_n$$

$$T_{DL} = \min_{n}\ \left(FDL\left(\hat{\boldsymbol{O}}_n, c\right)\right)$$

(14.4)

CONSTRAINED AUTONOMY

The concept of constrained autonomy for ships was proposed by Rødseth and Nordahl (2017). Here, it was defined as the automation system having defined limits to the options it can use to address these conditions, e.g. maximum deviation from planned track or arrival time. The automation system needs to request assistance from human operators if these constraints are exceeded. This definition was linked to the concept of operational envelope in (Rødseth 2019). The latter paper also defines five degrees of automation:

- *DA0 – Operator controlled*: Limited automation and decision support is available, as on most of today's merchant ship. The human is always in charge of operations and need to be present at controls and aware of the situation at all times.
- *DA1 – Automatic*: More advanced automation, e.g. dynamic positioning, automatic crossing or auto-berthing, is used. Crew attention is required to handle problems such as object classification and collision avoidance. The human may use own judgement as to how long he or she may be away from the control position. For automated fjord crossing in good weather, little traffic and in sheltered water, the operator may be away from the controls for several minutes.
- *DA2 – High automation*: The degree of automation is higher than for DA1 and may include certain "cognitive" functions, such as object detection and classification or collision avoidance. However, there are inherent and unknown limits to the automation system's capabilities.. These limits are not defined or constrained (see DA3), so the human operator must still use his or her judgement as to the required attention level. However, it is assumed that the need for attention is lower than for DA1.
- *DA3 – Constrained autonomy*: The degree of automation is similar to DA2, but system capabilities are now constrained by programmed or otherwise defined limits. The limits are set to enable the system to detect when limits are exceeded and to alert the operator in time before operator intervention is required.

- *DA4 – Full autonomy*: The ship automation can handle all states in O without any intervention from a crew.

These levels are related to the concepts of T_{DL} and T_{MR} and one can infer a link to the abstraction level of the crew–automation interface. This is shown in Table 14.1 where the automation levels are listed together with other relevant parameters.

By using the definitions of the operational domain's division into spaces for human and automatic control, it is possible to define constrained autonomy by the following more formal requirements:

1. All system states c that can be reached when the system is constrained autonomous, i.e. in O_{AC}, must have a known response deadline.
2. For all states c in O_{AC}, the function to determine T_{DL} (FDL), when called after an interval Δt must always return a T_{DL} that is not shortened by more than Δt.
3. For all states c in O_{AC}, the operators must be alerted to take command, i.e. requested to intervene (RTI), when T_{DL} is reduced to T_{MRS}.

The above requirements can be formulated as in Eq. (14.5), where $c(t)$ means the system state at time t and $RTI()$ means activation of an RTI. This specifies the necessary requirements to constrained autonomy and the corresponding subdivision of the operational envelope O_{AC}.

$$\forall O_n \subset O_{AC}, \forall c \in O_n : \begin{pmatrix} FDL(O_n,c) > 0 \\ FDL(O_n,c(t)) - FDL(O_n,c(t+\Delta t)) < \Delta t \\ FDL(O_n,c) \leq T_{MRS} \Rightarrow RTI() \end{pmatrix} \quad (14.5)$$

The original definition of constrained autonomy (Rødseth & Nordahl 2017) only required constraints on the decision capabilities and the output of the automatic control functions. This was later supplemented by specifications of response deadlines and maximum response times (Rødseth 2019). This text has updated the definitions of the deadline and the response times as well as of the different control spaces of O.

TABLE 14.1
Five Degrees of Automation

Degree of Automation	T_{DL}	Abstraction Level	Crew Attendance
DA0: Operator controlled	0	Functional	Continuous
DA1: Automatic	?	Functional	By judgement
DA2: High automation	?	Goal	By judgement
DA3: Constrained autonomy	$>T_{MR}$	Goal	Periodically unattended
DA4: Full autonomy	∞	Goal	Uncrewed

This has made it possible to have a more formal definition of constrained autonomy as provided above.

SUMMARY AND CONCLUSIONS

The concept of constrained autonomy as it has been presented here has been developed over several years and is probably not fully completed yet. However, the concepts and definitions presented in this text are now being used in several development projects as it is expected that the basic ideas are reasonably stable. There may still be some changes in the details of the definitions and in how it will be implemented in actual automation systems. This will be reported on in future publications.

The initial proposal of this text was that a more deterministic automation system in industrial autonomous systems may increase the operator's trust in the automation and by that improve the efficacy of the HAI for periodically unattended autonomous operations. A central element is to have a verifiable response deadline that can be matched to the crew's maximum response time as determined by the organization of watch-keeping on the ship and in the RCC. The theory is that this is likely to help in alerting the crew in time to establish a sufficient situational awareness before the crew is forced to act on situations that the automation system is not able to handle itself. Thus, the use of constrained autonomy should be a useful way to let the operators make better sense of the interaction between constrained autonomous systems and the operators. However, the analysis of the human–automation effects is beyond the scope of this text and will have to be addressed by experts in the human factors field. While the initial proposal seems logical, it will be up to researcher in the area of human factors to see what actual implications constrained autonomy has on the operators' ability to make better sense of the interaction between human and automation

The main purpose of this text is therefore to describe the technical concept of constrained autonomy and give it a more formal definition. Some examples of consequences for implementations of automation systems have also been given.

Independent of the human-factor angle, it is also believed that the concept can be used to improve testability and eventually also formal acceptance of autonomous control systems (Rødseth 2019). The concept of constrained autonomy may be particularly important for industrial autonomous systems, where the systems are costly, need to operate in a commercial business model and where the consequences of system failures may have significant and even catastrophic consequences. Industrial autonomous systems are also very relevant for the concept of constrained autonomy as many of them will need an operator in the loop in any case, mainly to oversee the operation and to safeguard large investments. The examples in this text are from the maritime domain and the work presented has been focusing on autonomous ships. However, the concept of constrained autonomy should also be applicable to other industrial autonomous systems.

The work presented in this text has been partially funded by the Norwegian Research Council project SAREPTA. It has also received funding from the European Union's Horizon 2020 research and innovation programme under grant agreement No 815012 (AUTOSHIP).

REFERENCES

Al-Rifaie, M. M., Bishop, J. M. & Caines, S. (2012). Creativity and autonomy in swarm intelligence systems. *Cognitive Computation, 4*(3), 320–331.

Grøtli, E., Reinen, T., Grythe, K., Transeth, A., Vagia, M., Bjerkeng, M., Rundtop, P., Svendsen, E., Rødseth, Ø. & Eidnes, G. (2015). Seatonomy design development and validation of marine autonomous systems and operations. In *Proc. MTS/IEEE OCEANS'15.* Washington, DC.

Hasegawa, K. (2004). Some recent developments of next generation's marine traffic systems. *IFAC Proceedings of Computer Applications in Marine Systems (CAMS'04)*, Ancona, Italy, 2004.

Hoff, K. A. & Bashir, M. (2015). Trust in automation: Integrating empirical evidence on factors that influence trust. *Human Factors, 57*(3), 407–434.

ISO (2020a). *Input Document to Maritime Safety Committee, Session 102, Agenda Item 5: Regulatory Scoping Exercise for the Use of Maritime Autonomous Surface Ships (MASS), Proposed terminology for MASS*, Submitted by International Organization for Standardization (ISO), February 2020.

ISO 23860 (2020b). *Ships and marine technology — Terminology related to automation of Maritime Autonomous Surface Ships (MASS)* – Internal Committee Working Draft 2.

Jha, S., Raman, V., Sadigh, D. & Seshia, S. A. (2018). Safe autonomy under perception uncertainty using chance-constrained temporal logic. *Journal of Automated Reasoning, 60*(1), 43–62.

Klein, G., Moon, B. & Hoffman, R. R. (2006). Making sense of sensemaking 1: Alternative perspectives. *IEEE Intelligent Systems, 21*(4), 70–73.

Lee, J. D. & See, K. A. (2004). Trust in automation: Designing for appropriate reliance. *Human Factors, 46*(1), 50–80.

Rødseth, Ø. J. & Burmeister, H. C. (2012). Developments toward the unmanned ship. In *Proceedings of International Symposium Information on Ships–ISIS*. Vol. 201.

Rødseth, Ø. J. & Nordahl, H. (2017). Ed. *Definition for autonomous merchant ships*. Version 1.0, October 10, 2017. Norwegian Forum for Autonomous Ships. Retrieved February 2020 from http://nfas.autonomous-ship.org/resources-en.html

Rødseth, Ø. J. (2018). Assessing business cases for autonomous and unmanned ships. In *Technology and Science for the Ships of the Future. Proceedings of NAV 2018: 19th International Conference on Ship & Maritime Research*. IOS Press.

Rødseth, Ø. J. (2019). Defining ship autonomy by characteristic factors. In *Proceedings of the 1st International Conference on Maritime Autonomous Surface Ships*. SINTEF Academic Press.

SAE J3016 (2016). *Taxonomy and definitions for terms related to on-road motor vehicle automated driving systems*. Revision September 2016, SAE International.

Tass (2018). *S7 Space to modernize Sea Launch floating spaceport for reusable rocket*. Retrieved February 2020 from https://tass.com/science/1029619.

Vagia, M., Transeth, A. A. & Fjerdingen, S. A. (2016). A literature review on the levels of automation during the years. What are the different taxonomies that have been proposed? *Applied Ergonomics*, 53, 190–202.

Yara, A. S. (2018). *The first ever zero emission, autonomous ship*. Retrieved February 2020 from https://www.yara.com/knowledge-grows/game-changer-for-the-environment/.

15 HMI Measures for Improved Sensemaking in Dynamic Positioning Operations

L. Hurlen
IFE

CONTENTS

INTRODUCTION

There is currently a considerable drive toward increasing levels of automation across all complex, safety-critical industries. In the transportation domain, we see numerous autonomous concepts of operation emerging, such as self-driving cars, autonomous ships, remotely operated drones and underwater vehicles – all concepts where the role of the operator is shifting from active, hands-on operation toward a more managerially oriented role focused on administering and supervising a suite of automated systems. In commercial aviation, this has been especially prevalent and has contributed to an impressive safety trend over the past few decades, but we have also seen new types of accidents resulting from a breakdown in the collaboration between the pilots and various automatic control systems. A recent example is the two Boeing 737 MAX 8 crashes where a newly installed automatic anti-stall system unexpectedly and repeatedly pushed the nose of the plane down. To make matters worse, it was designed in a way that made manual intervention difficult even when the pilots finally understood what was happening (National Transportation Safety Board, 2019).

In this chapter, we look at this challenge from the perspective of a modern ship bridge, and more specifically the operation of dynamic positioning systems (DP systems). This technology is utilized for automatic station keeping and is becoming ubiquitous in the maritime domain in a wide variety of operations such as drilling, cargo loading, diving operations and pipe-laying. As such, various industries increasingly depend on the safe operation of these systems, and the considerable number of incidents and accidents that have occurred in recent years are causing concern.

The work presented in the following is the result of a study performed within the sensemaking in safety-critical situations (SMACS) research project (SINTEF, 2018) supported by the Norwegian Research Council and industry partners Human Factors in Control (HFC) forum and Kongsberg Maritime, focusing on human–machine interfaces (HMIs). Sensemaking refers to the ability of operators to perceive and understand situations and act accordingly in complex environments (Kilskar et al., 2018; Weick, 1988). It is closely related to "Situation Awareness" (SA) as described by Endsley and Jones (2012). The purpose of this study has been to identify the factors that challenge the sensemaking of DP operators (DPOs) and to propose new design principles for effective human–automation interaction that may improve safety.

This has been a mixed-method feasibility study with a strong focus on end-user involvement and learning from related safety-critical domains that Institute for Energy Technology (IFE) has worked with. The study first identified key challenges through semi-structured interviews with instructors and experienced DPOs, observations during simulator-based DP training, discussions with Equinor "Captains forum" and analysis of incident and accident reports, summarized in Hurlen, Skjerve & Bye (2019). Design opportunities was then explored and exemplified through mock-ups (Hurlen & Bye, 2020) and finally evaluated with end-users, summarized at the end of this chapter.

DYNAMIC POSITIONING (DP)

To understand the context, let us first look at DP operations, how the system works and how it is used. The system itself works by automatically controlling thrusters and rudders to keep a predetermined position, using input from a variety of reference systems – position reference systems (such as radar, GPS, hydroacoustic and laser systems) and sensors measuring external forces acting on the vessel (wind and current) – to compute and execute the force necessary for station-keeping. DP systems are classed 1–3 according to their level of technical redundancy. Class 3 is generally required for safety-critical operations and involves the capability of no single fault in an active system causing the system to fail and is also being able to withstand fire or flood in any one compartment without the system failing (see IMO publication 645).

For many types of operations, this is increasingly becoming a preferred way of maintaining position as an alternative to anchoring. Depending on the operation, DP systems are used more or less prevalently. A cargo vessel may use it for only a few

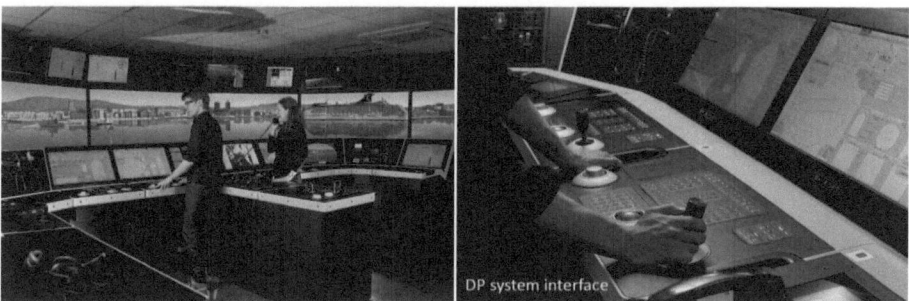

FIGURE 15.1 Modern ship bridge environment with dynamic positioning (DP) systems. (Training simulator, photo by Kongsberg Maritime.)

hours during offloading while a drilling or construction vessel may spend most of its time on DP. While the vessel is on DP, a dedicated DPO on the ship bridge is responsible for supervising and controlling the system, which includes ongoing risk assessment, setting and adjusting position setpoints according to changing operational needs and weather conditions, managing reference systems and collaborate with the rest of the ship crew. The DP system can operate in a variety of modes, including full auto position and a variety of combinational modes where the DPO assists the automation in specific ways. For example, the system can be set to automatically maintain the heading of the vessel into the prevailing weather (weathervane mode), follow a remotely operated underwater vehicle (ROV) or maintain position less strictly in order to save fuel (eco mode).

Figure 15.1 shows a modern ship bridge with DP interfaces. As we can see, this is a highly computerized environment. Although a typical bridge still features a lot of analogue equipment, digitalization are a major trend. Computer graphics is increasingly being utilized to convey information about the vessel, its control systems and its surroundings. As shown in Figure 15.1, DP system interfaces typically consist of a panel with physical buttons and joysticks for key mode selections and system configurations, and a number of configurable screens that show system status, alarms, trends and other relevant information. Advanced diagnostic features include a "capability plot" that graphically draws a boundary around the vessel representation, indicating available directional force in case of worst case single-failure event and consequence analysis warnings.

CURRENT CHALLENGES

Accidents and near-misses reports indicate that the sensemaking of DPOs is not always successful. As in most other complex, fast-paced, safety-critical environments, DPOs often face significant challenges when trying to get a proper overview and make sense of the situations they find themselves in. As control systems are increasingly being digitalized, the available amount of information and control opportunities escalate, and the way they are designed most often leaves the human operator with the task of navigating and counteracting their various strengths and weaknesses, capabilities

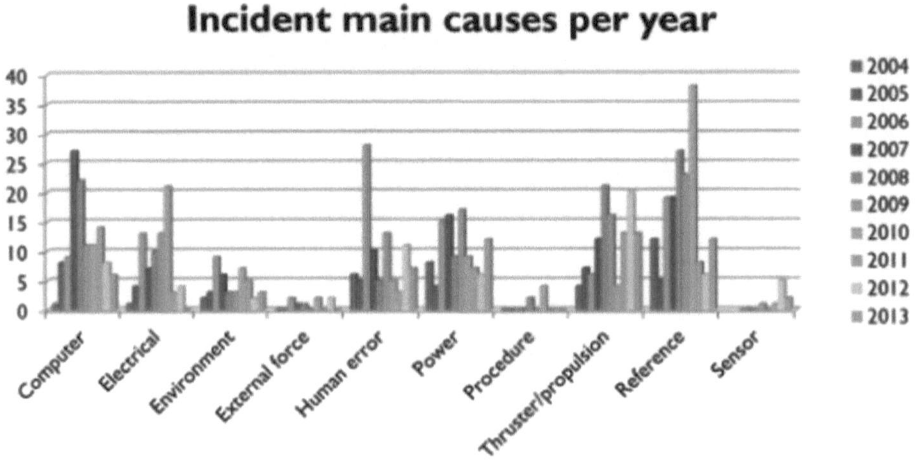

Incident main causes per year

FIGURE 15.2 Dynamic positioning (DP)-related incidents reported to IMCA, 2004–2013.

and limitations. From a purely technical perspective, much of current ship navigation and control system management is fully automated, but the operators are ultimately responsible for maintaining safety – the final barrier in the chain of defense. In the years 2004–2013, up to 27 DP-related yearly incidents reported to IMCA were labeled "human error" as their root cause, see Figure 15.2 (IMCA, 2016).

Our analysis suggests that these human errors most often can be traced back to poor system design – to putting human operators in a position where they are unreasonably vulnerable to failure. In this study, we therefore adopt a user-centered design perspective, assuming that when technology is properly designed and aligned with human capabilities and limitations, operators are able to work safely and effectively in collaboration with it. Based on interviews, observations and incident report analysis, we have identified six main sensemaking-related challenges that are currently facing DPOs (Hurlen, Skjerve & Bye, 2019): (1) alarms, (2) mode surprises, (3) critical information hidden from view, (4) "Private" HMIs limits shared SA, (5) deskilling and (6) out-of-the loop.

Alarms seem to be an issue that is particularly challenging for DPOs, either because there are too many alarms being announced in a short period of time, or that alarms or warnings that should have been announced were not, or that they were not properly recognized or clearly understood. Another category that we found particularly interesting because of its familiarity with a well-known issue in related safety-critical domains is the "critical information hidden from view" challenge. DPOs require an extensive amount of status information from the DP system in order to maintain their SA. Because of space limitations on the bridge, a typical DP setup consists of one to three screens. Since not all relevant information can be presented simultaneously on these screens, system providers allow users to organize the content and layout quite freely – so-called user-configurable screens. And since safety-critical situations often happen rapidly and unexpectedly, the information that is presented on user-configurable screens at any one moment may or may not be the one the DPOs need to make a correct assessment of the situation and to best evaluate the effect of any countermeasures that are being made. It could be that the current user has actively organized the screens to best fit the planned situation (and thus not necessarily for the developing off-normal situation) or that the previous user has arranged the screens to fit his or hers personal preferences which may not entirely serve the current user and situation. But since the screens can be re-configured at any time by the users, why is this a problem? Why cannot the DPOs simply reorganize their screens to best fit changing circumstances?

When the nuclear industry began digitalizing powerplant control rooms, similar issues were raised. In the old, analogue control rooms, every process function and every component had its own dedicated, analogue control mechanism, located at a fixed position in the room (so-called "spatially dedicated" interfaces). The new screen-based control systems on the other hand were highly flexible (user configurable), and the users could bring up any component or any piece of information on any screen. By system developers this was considered an advantage: more compact control environments could be made, more powerful information graphics could be designed, control functions could be organized according to changing circumstances and needs, interface maintenance could be performed with less effort, etc. All good stuff, but could all this flexibility also lead to confusion and reduced process overview? Operators were concerned. A study was initiated by the US Nuclear Regulatory Commission (NRC, 2002) to investigate the effects of interface management tasks on operator performance. It concluded:

> When HSIs [Human System Interfaces] are spatially dedicated, operators can use automatic information processing capabilities, such as scanning and pattern recognition, to rapidly assess plant situations. The flexibility of computer-based HSIs and their general lack of spatial dedication causes interface management tasks to be more dependent on controlled information processing. The flexibility also makes it easier for operators to mistake one display for another, and may cause them to improperly assess a situation or operate the wrong piece of equipment.
>
> *(O'Hara et al., 2002, p.7)*

The study also noted that operators were less likely to perform interface management tasks during stressful situations, relying instead on information that were immediately available to them. That operators coming from an analogue control room to a digitalized one often find it difficult to get a proper situation overview is supported by the research performed at IFE Halden for the nuclear industry (e.g. Kaarstad & Strand, 2010; Kaarstad et al., 2008). This helps explain why "information hidden from view" is indeed a challenge for DPOs. During our interviews, we noted several statements related to this. One said: "He who sat there [at the DP desk] before might think he is the world champion and has changed everything around. Very many might then miss important information, especially signals that point towards things that can go wrong." Another stated that "if the information is not already on the screen it will not be used".

This point toward a potentially effective design measure for digital control environments that is increasingly becoming industry standard in the nuclear and petroleum industries: The "spatially dedicated" overview display.

OVERVIEW DISPLAYS FOR IMPROVED SITUATION AWARENESS

Based on studies like the ones mentioned in the previous section, IFE has worked extensively with the petroleum and nuclear industry to explore concepts and solutions that would take advantage of new possibilities as well as mitigate some of the challenges associated with the emerging computerized interfaces, see Braseth et al. (2009) for a summary. Among the most influential concepts that has been developed is the shared overview display, sometimes also referred to as a Large Overview Display or Group-View Display. The purpose of an overview display is to provide individuals and teams in the control room with an at-a-glance overview of the most important safety-critical process parameters at a fixed location in the room, enabling them to quickly assess the current situation, notice deviations early and to prioritize effectively between multiple events without having to "dig" for information in the control system (navigate). After a series of lab studies, industry development projects and experience reviews, a growing body of knowledge attests to their effectiveness (see Laarni et al., 2009; Hurlen et al., 2015; Kortschot et al., 2018; Kaarstad & Strand, 2011; Roth et al., 2001; Veland et al., 2010). When combined with fully flexible interfaces, a well-designed overview display enables the operators *simultaneously* to adapt the HMI to changing circumstances *and* to quickly assess the "big picture" without missing important information. Figure 15.3 shows a typical petroleum control room with a shared overview display.

An emphasis in IFEs overview display design work has been given to designing effective information graphics – visuals that enable operators to directly *see* the process status rather than having to read/reflect on e.g. textual alarm descriptions, providing faster decision support that is especially valuable in stressful situations. As an example, IFE has patented the "mini trend" graphic (Braseth, 2015; adapted for nuclear domain in Svengren et al., 2014, p. 153) illustrated in Figure 15.4. This graphic combines the trend line, relevant alarm limits and the current numerical value in a timeframe suited for drawing operators' attention early to developing deviations.

FIGURE 15.3   Petroleum control room combining a shared overview display (top) and flexible screens (bottom).

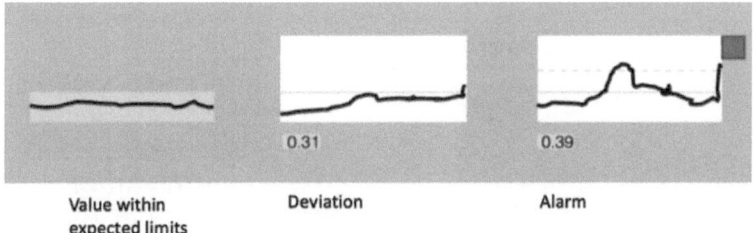

FIGURE 15.4 Mini-trend information graphic.

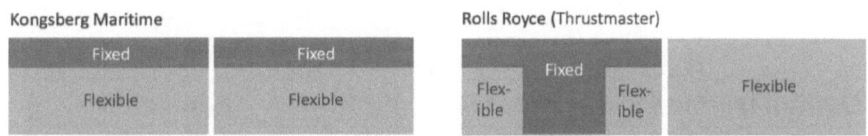

FIGURE 15.5 Dynamic positioning (DP) screen layout: two current industry practice examples that combine fixed and flexible content.

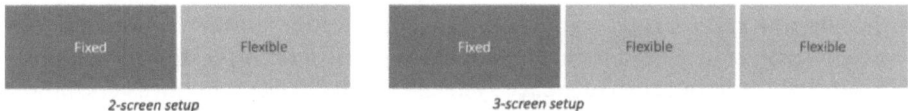

FIGURE 15.6 Alternative layout dedicating more screen space to fixed content.

TOWARD AN OVERVIEW DISPLAY DESIGN FOR DP OPERATION

Could a fixed (spatially dedicated) overview display also help DPOs strengthen their SA? Current industry DP systems also combine flexible and fixed elements in their layouts. Figure 15.5 shows common DP system HMIs, illustrated in a two-screen setup.

Based on the "critical information hidden from view" challenge that has been identified, we wanted to investigate whether SA could be improved by extending the screen area dedicated to fixed information. Figure 15.6 shows an alternative layout design.

To assess the feasibility and effectiveness of this idea, three main questions need to be answered: (1) Is it possible to define a unique set of safety-critical information elements that provides DPOs with "the big picture" relevant in all (or at least the most safety-critical) situations? (2) Is it possible to present this information on a display that is compact enough to fit within the DPOs field of view on typical bridge environments? (3) Could such a design improve DPO performance (sensemaking and SA) in safety-critical situations?

In Hurlen & Bye (2020), we explored a possible design in order to exemplify how an effective solution might be accomplished and made mock-ups suited for

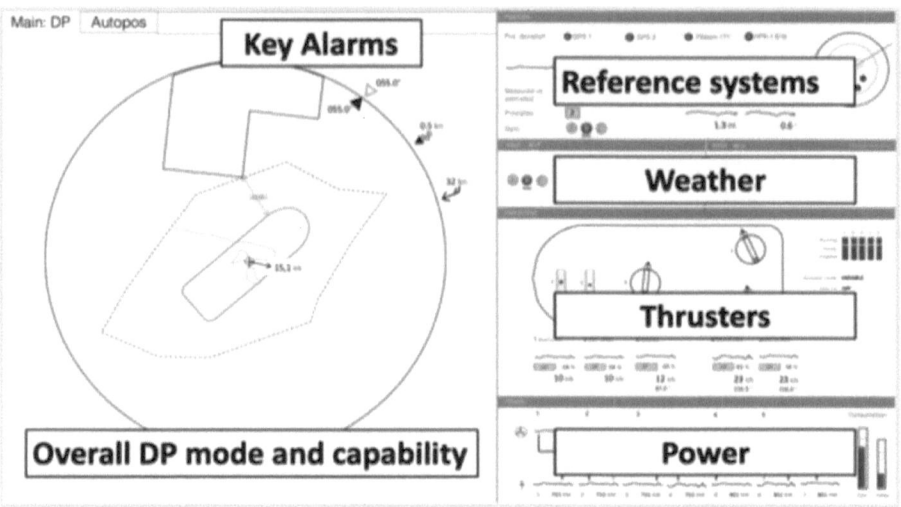

FIGURE 15.7   Possible overview display design layout and main content. (From Hurlen & Bye 2020.)

evaluation with end-users based on the information needs that were identified so far. The main content includes absolute and relative position of the vessel, current position setpoint, status of active reference systems, weather conditions, thrusters, power supply systems and alarms/warnings. A possible layout for a DP overview display is shown in Figure 15.7.

Based on the lessons learned from overview display design in the nuclear and petroleum domain, mock-ups were made to illustrate how the display could behave during different circumstances. Key design objectives in this work were to support "at-a-glance" use, creating information graphics that might give DPOs early warnings and thus extended time to handle disturbances before they reach a critical stage, and highlighting automation mode-related information for which our interviews and incident analysis also identified some challenges.

Please remember that this design is meant to be combined with flexible elements needed for necessary HMI adaption to changing circumstances. The design mock-ups are shown in Figures 15.8–15.10.

RESULTS FROM EVALUATION WITH END-USERS

To assess the feasibility of the overview display design idea, the mock-ups described above were presented and discussed with four experienced DPOs. They each had 3–9 years of operative DPO experience from a variety of vessels and operations, including cargo, construction and production vessels (rigs). Two were also experienced DP instructors. The DPOs were interviewed individually on video. The interviews were performed in a semi-structured manner and organized around the four main topics presented below. A possible weakness with this method is that interview

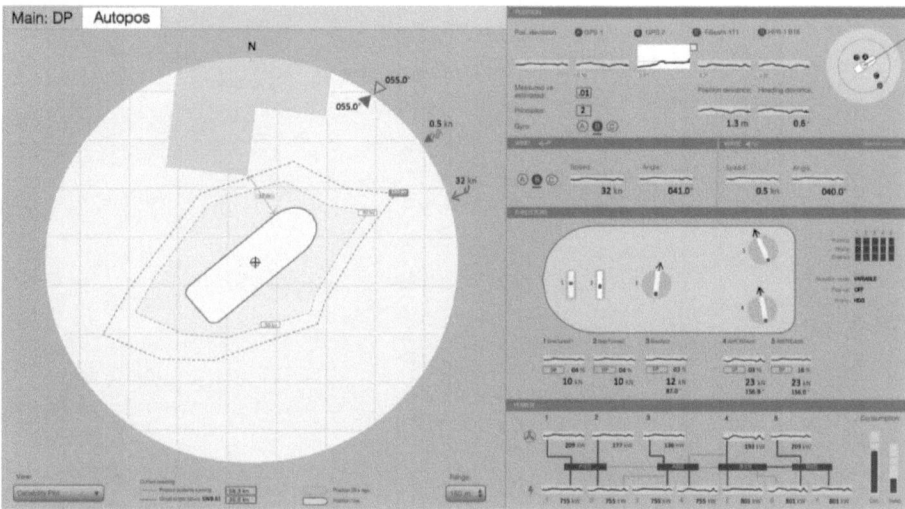

FIGURE 15.8 Possible overview display design – normal operation with minor disturbance. (From Hurlen & Bye 2020.)

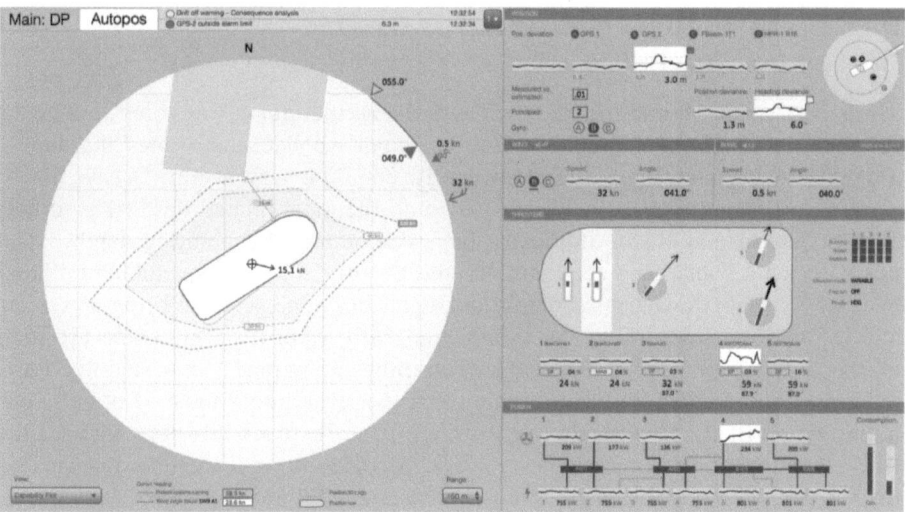

FIGURE 15.9 Possible overview display design – off-normal situation with multiple disturbances. (From Hurlen & Bye 2020.)

subjects may be inclined toward agreeing or being overly positive to the ideas presented to them. We sought to counteract this effect by encouraging subjects to freely express any views they had and by asking them to elaborate on their input as well as give practical examples from their own experience, which all of them did. No compensation was offered for their participation.

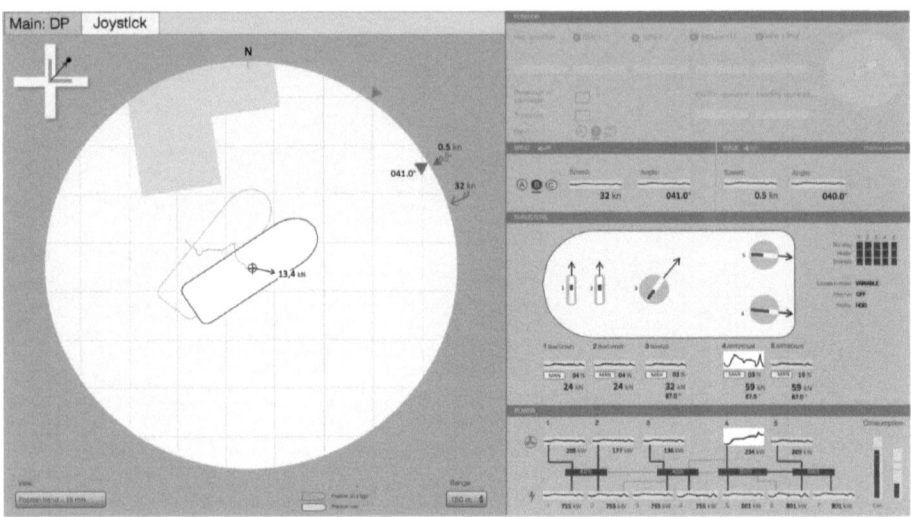

FIGURE 15.10 Possible overview display design – autopos mode is disengaged (return to joystick mode). (From Hurlen & Bye 2020.)

FEASIBILITY OF THE FIXED OVERVIEW DISPLAY CONCEPT – DOES THE IDEA SEEM PROMISING?

Three participants found the idea very promising, the fourth was somewhat doubtful about its potential for success. On the supportive side, they stated that having a fixed screen giving them a broad overview could be very useful: "I have always wanted something like this". "What would do the most related to HMI is to have an overview display that is intuitive, others can be used for going in-depth". "One place to cast a glance and see that everything is ok." The content should be fixed, because "…when the alarms screams and it gets busy one needs to bring up screens and information that covers the things you really want [to see]. There are more alarms during bad weather and that's when things go wrong". They said that introducing a fixed display with information that is useful across most, if not all, situations would likely be possible to realize and that the mock-ups presented seemed like a good starting point for design. The participants thought that an overview display would be possible to fit in most current vessel types, but more easily on rigs than, e.g., supply vessels because of available space limitations. One stated that on a rig he would gladly dedicate two 50 in. displays for overview information in addition to the three 27-in. screens he is currently using for DP. Another pointed to the paper-based checklist that DPOs go through when starting their shift as a promising starting point for content selection, stating that "if I had a setup with all the things in the checklist I would always have it up." Some speculated that an overview display might be useful not only for the DPO but for the rest of the bridge crew as well as for visitors (a possible measure for addressing the "Private HMIs limits shared situation awareness" challenge mentioned earlier). In this case, one might consider including non-DP related content also to create a more comprehensive overview for several bridge systems. This would need to be aligned with the needs of the DPO, possibly duplicating the display in several locations. The concerns were mainly related to the fact that information needs vary across circumstances, thus the DPOs need to be able to reorganize their screens. One of the interviewees felt that if there is an inexperienced DPO on duty, the captain should take an active role in dictating HMI layout arrangements.

CRITICAL SUCCESS FACTORS

Participants agreed that the design should be clear and as simple as possible, with great emphasis on early warnings and deviations from expected normal states. The display should not feel cluttered, there might be a pitfall to include too much content that is "nice to have".

IDEAS FOR IMPROVING THE SKETCHES PRESENTED

One commented that perhaps wind presentation could be displayed to more clearly announce changes in strength and/or direction, e.g. could the wind arrows change color to yellow when it is increasing. Many lower level components could perhaps be presented using a collective "traffic light" to enrich display content while saving space. Using trended information as extensively as in the sketches seem promising, but they should not be too small – the alarms should be more clear. One speculated that maybe it would be a good idea if some information could be brought up by the user based on the situation, such as the capability plot (the sketches in Figures 15. 8–15.10 actually indicates the possibility of selecting among predefined "views", located at the bottom right). If a vessel is on Posmoor(atar) – a combination of DP- and anchor-based position control – more information related to this should be found on the display.

OTHER INPUT OR SUGGESTIONS

All participants stressed the importance of improving the alarm situation. An overview display should acknowledge this challenge, aiming in its design to further help DPOs notice and correctly diagnose various warnings and alarm situations. Several of the participants were of the impression that DP is utilized more and more in the petroleum domain. Some also expressed concern about management (on shore) pushing operations too far, reducing the safety margins in order to maximize profits.

CONCLUSION AND FURTHER WORK

By analyzing DP-related incidents and interviewing experienced DPOs, captains and instructors, we have found that there is a potential for making changes to the HMI design in order to improve sensemaking. We have found that the challenges DPOs face are similar to the ones operators in other safety-critical industries are experiencing, so there are apparent opportunities for the maritime domain to learn from how they are being addressed elsewhere. In this work, we have looked to HMI and control room design in the petroleum and nuclear domains in particular for inspiration and propose introducing a fixed overview display as a supplement to existing HMIs as a feasible design measure for improving SA and sensemaking for DPOs. Current DP systems are typically designed in a way that let operators arrange their screens quite freely, leaving them vulnerable to overlook important information when unexpected situations occur – which they sometimes do and often quite rapidly. This study indicates that to introduce a well-designed fixed overview display can give DPOs an at-a-glance system overview regardless of the situation, enabling them to quickly detect, understand and counteract developing off-normal situations. On existing vessels, it might be more feasible to introduce such displays on construction and productions vessels (rigs) than on e.g. cargo vessels due to space limitations.

This has been an exploratory study with relatively few user participants, thus findings are far from conclusive. Still, the results provide motivation for DP systems developers to question one of the fundamental design choices related to HMI design – display reconfigurability versus fixed content presentation. The design proposals presented in this study are likely a good starting point for further user- and performance-driven research and development.

ACKNOWLEDGMENTS

Thanks to Andreas Bye for helping out with the evaluation interviews and providing useful feedback, to Kongsberg Maritime – in particular Jan Inge Edvardsen – for their feedback and for providing access to their systems and simulators, and to all the DPOs that participated.

REFERENCES

Braseth, A. O. (2015). *Information-Rich Design: A Concept for Large-Screen Display Graphics: Design Principles and Graphic Elements for Real-World Complex Processes*, Doctoral thesis at NTNU; 2015: 30.

Braseth, A. O., Nihlwing, C., Svengren, H., Veland, Ø., Hurlen, L. and Kvalem, J. (2009). Lessons learned from Halden project research on human system interfaces. *Nuclear Engineering and Technology* 41(3). DOI: 10.5516/NET.2009.41.3.215

Endsley, M. R. and Jones, D. G. (2012). *Designing for Situation Awareness, An Approach to User-Centered Design*. Second Edition. CRC Press.

Hurlen, L. and Bye A. (2020). Improving sensemaking in dynamic positioning operations: HMI and training measures. In *Proceedings of the 30th European Safety and Reliability Conference*.

Hurlen, L., Skjerve, A.B. and Bye, A. (2019). Sensemaking in high-risk situations. The challenges faced by dynamic positioning operators. In *Proceedings of the 29th European Safety and Reliability Conference*.

Hurlen, L., Skraaning, G., Meyers, B., Carlsson, H. and Jamieson, G. (2015). The plant panel: feasibility study of an interactive large screen concept for process monitoring and operation. In *9th International Topical Meeting on Nuclear Plant Instrumentation, Control, and Human Machine Interface Technologies (NPIC&HMIT 2015)*.

IMCA (2016). *IMCA M 166 Rev. 1 – Appendix 3 – The IMCA DP Station Keeping Incident Database*. International Marine Contractors Association.

IMO Publication 645, https://www.kongsberg.com/maritime/support/themes/imo-dp-classification/).

Kaarstad, M. and Strand, S. (2010). *Work practices: field study of challenges and opportunities in a computer-based nuclear powerplant control room.* OECD Halden Reactor Project work report HWR-1053 Rev2.

Kaarstad, M. and Strand, S. (2011). *Large screen displays—a usability study of three different designs.* OECD Halden Reactor Project report HWR-1025.

Kaarstad, M., Strand, S. and Nihlwing, C. (2008). *Work practices in computer-based control rooms – insights from small-scale studies with operators.* OECD Halden Reactor Project work report HWR-892 Rev2.

Kilskar, S.S., Danielsen, B.E. and Johnsen, S.O. (2018). Sensemaking in critical situations and in relation to resilience—a review. *ASCE-ASME Journal of Risk and Uncertainty in Engineering Systems, Part B: Mechanical Engineering* September 2019, DOI: https://doi.org/10.1115/1.4044789.

Kortschot, S., Jamieson, G. and Wheeler, C. (2018). Efficacy of group-view displays in nuclear control rooms. *IEEE Transactions on Human-Machine Systems* PP(99):1–7.

Laarni, J., Koskinen, H., Salo, L., Norros, L., Braseth, A.-O. and Nurmilaukas, V. (2009). Evaluation of the Fortum IRD pilot. *Paper presented at the Sixth American Nuclear Society International Topical Meeting on Nuclear Plant Instrumentation, Control, and Human-Machine Interface Technologies, NPIC&HMIT 2009,* Knoxville, Tennessee, April 5–9, 2009, American Nuclear Society, LaGrange Park, IL.

National Transportation Safety Board (2019). *Safety recommendation report. Assumptions used in the safety assessment process and the effects of multiple alerts and indications on pilot performance.* Accident number: DCA19RA017/DCA19RA101.

O'Hara, M., Brown, W.S., Lewis, P.M. and Persensky, J. (2002). *The effects of interface management tasks on crew performance and safety in complex, computer-based systems.* US Nuclear regulatory Commission. NUREG/CR-6690, vol. 1/BNL-NUREG-52656, vol. 1.

Roth, E.M., Lin, L., Kerch, S., Kenney, S.J. and Sugibayashi, N. (2001). Designing a first-of-a-kind group view display for team decision making: a case study. In Salas, E. and Klein, G. (ed) *Linking Expertise and Natural Decision Making.* Lawrence Erlbaum.

SINTEF (2018). SMACS project website, https://www.sintef.no/projectweb/hfc/smacs/

Svengren, H., Hurlen, L. and Nihlwing, C. (2014). Human machine interface (HMI) developments in HAMMLAB. *International Electronic Journal of Nuclear Safety and Simulation* 5(2), 149–158.

U.S. Nuclear Regulatory Commission (2002). Human system interface design review guideline. Revision 2, section 6: Group-view display system. U.S. Nuclear Regulatory Commission Office of Nuclear Regulation (NUREG-0700 Rev. 2)

Veland, Ø., Eikås, M., Andresen, G., Hurlen, L., Weyer, U. and Kristiansen, P. (2010). *Design patterns for large screen displays – lessons learned from the petroleum industry, HWR-933.*

Weick, K.E. (1988). *Enacted sensemaking in crisis situations. Journal of Management Studies,* 25(4), 305–317.

Index

Note: **Bold** page numbers refer to tables and *italic* page numbers refer to figures